Aquaculture: An Introductory Text

Aquaculture: An Introductory Text

Diana Fletcher

STATES
ACADEMIC PRESS
www.statesacademicpress.com

Aquaculture: An Introductory Text
Diana Fletcher
ISBN: 978-1-63989-054-5 (Hardback)

Published by States Academic Press,
109 South 5th Street,
Brooklyn, NY 11249, USA

Cataloging-in-Publication Data

Aquaculture : an introductory text / Diana Fletcher.
p. cm.
Includes bibliographical references and index.
ISBN 978-1-63989-054-5
1. Aquaculture. 2. Agriculture. I. Fletcher, Diana.
SH135 .A68 2022
639.8--dc23

Table of Contents

Preface

The process of rearing, breeding and harvesting of both aquatic animals and plants in controlled environments like rivers, lakes, oceans, ponds and streams is known as aquaculture. The practice of aquaculture serves various purposes such as food production, restoration of endangered species, habitat restoration, creation of barrier against pollution using seaweed and mollusc, and wild stock population enhancement. Aquaculture can be practiced in both saltwater and freshwater. Common aquaculture practices include mariculture, algaculture, inland pond culture and integrated multi-trophic aquaculture. Some common fish that are bred using aquaculture techniques are catfish, salmon, carp, shrimp, oyster and tilapia. This book elucidates the concepts and innovative models around prospective developments with respect to aquaculture. It is compiled in such a manner, that it will provide in-depth knowledge about the theory and practice of aquaculture. Coherent flow of topics, student-friendly language and extensive use of examples make this book an invaluable source of knowledge.

A detailed account of the significant topics covered in this book is provided below:

Chapter 1- Aquaculture or aqua farming is the breeding, rearing and harvesting of aquatic organisms like fish, shellfish, prawn, shrimp, etc. It can be broadly defined as the farming of both animals and plants in freshwater as well as saline waters. This chapter's aim is to acquaint the reader with aquaculture and its sustainable practices.

Chapter 2- Aquaculture comprises numerous methods and practices based on what, where and how it is practiced. Mariculture, algaculture, pond culture, running water culture, recirculating aquaculture systems, rise-fish culture, raceway aquaculture, monoculture, etc. are just a few types of aquaculture. The topics elaborated in this chapter will help in gaining a better perspective about the different methods and practices associated with aquaculture.

Chapter 3- Fish farming is a form of aquaculture in which fishes are raised in tanks or other similar enclosures for personal use or profit. Extensive aquaculture and intensive aquaculture are two basic types of fish farming. The occupation or enterprise of undertaking such an activity is known as a fishery. This chapter deals with fish farming and fisheries and the various types, practices and the issues plaguing them in a detailed manner for the reader's benefit.

Chapter 4- Like any other commercial activity, aquaculture has its own challenges. The complications brought about by unwanted parasites and diseases associated with fishes can have devastating effects on aquaculture. This chapter discusses the different diseases and parasites which can harm fishes, with special attention being given to the salmon.

Chapter 5- Aquaponics is a form of aquaculture in which the nutrient rich aquaculture water is fed to plants grown in a soil-less environment. It can be considered as a combination of aquaculture and hydroponics. Algaeponics, vermiponics, aquaponic systems, aquaponic fish, aquaponic vegetables, bio filters, etc. are some of the aspects related to aquaponics which will be explained elaborately in this chapter.

It gives me an immense pleasure to thank our entire team for their efforts. Finally in the end, I would like to thank my family and colleagues who have been a great source of inspiration and support.

Diana Fletcher

1

Understanding Aquaculture

Aquaculture or aqua farming is the breeding, rearing and harvesting of aquatic organisms like fish, shellfish, prawn, shrimp, etc. It can be broadly defined as the farming of both animals and plants in freshwater as well as saline waters. This chapter's aim is to acquaint the reader with aquaculture and its sustainable practices.

Aquaculture is the aquatic counterpart of agriculture, has grown rapidly in recent decades, and today it produces almost as much fish and shellfish as fisheries. Aquaculture is the main means for obtaining more food from our aquatic environments in the future. Impacts of aquaculture on biodiversity arise from the consumption of resources, such as land (or space), water, seed, and feed, their transformation into products valued by society, and the subsequent release into the environment of greenhouse gases and wastes from uneaten food, fecal, and urinary products, chemotherapeutants as well as microorganisms, parasites, and feral animals. Negative effects may be direct, through release of eutrophicating substances, toxic chemicals, the transfer of pathogens diseases and parasites to wild populations stock, and the introduction of exotic and genetic material into the environment, or indirect through loss of habitat and niche space and changes in food webs. Today, large quantities of fish are caught to produce fishmeal and fish oil – an important source of ingredients for protein and fatty acids in feeds for many fish and shrimp species. The volume for non-food usage is slowly shrinking but the still high demand from the aquaculture industry may lead to unsustainable fishing practices with potential negative consequences for fish stocks and marine food webs. Despite advances being made within the feed industry, resulting in decreased feed conversion ratios and development of suitable alternatives to fish resources, the aquaculture industry's use of global fishmeal and fish oil increased threefold between 1992 and 2006.

The life cycles for most aquaculture species have been successfully closed but the culture of some, especially marine, farmed fish, and shrimp, is still partially dependent on the capture of larvae, post larvae, or gravid females from the wild. This can result in both over-fishing and bycatch, representing losses to capture fisheries and biodiversity. Large areas of critical habitats such as wetlands and mangroves have historically been lost due to the siting of aquaculture developments and pollution, resulting in reduced biodiversity and recruitment to capture fisheries. Today the values of coastal wetlands are better recognized and regulations make it difficult for large-scale aquaculture development in sensitive coastal areas (e.g., mangroves, sea grass beds) in most countries.

The magnitude of biodiversity loss from aquaculture development generally increases with scale, intensity of resource use, and net production of wastes. In some cases aquaculture may also increase local biodiversity, for example, when ponds are constructed in dry areas, through re-stocking activities and from integrated aquaculture. Its role in maintaining cultural diversity must also be acknowledged.

Aquaculture Development and Practices

The farming of aquatic plants and animals is several thousands of years old. In 1950, global farmed fish and shellfish production was approximately 2 mmt (million metric tons) and largely confined to areas of Asia. During the last three decades, aquaculture production has increased by approximately 7–11% per year. Aquaculture production for 2012 was approximately 44 mmt of finfish, 6 mmt of crustaceans, 15 mmt of mollusks, and 23.8 mmt of aquatic plants. Fish produced by aquaculture now accounts for half of all fish directly consumed by humans although the bulk of this production is freshwater fish (64%, mainly carps) and mollusks (22%, oysters dominating). Around 600 different freshwater and marine animal species, representing different trophic levels, are cultured using a wide range of technologies and inputs. However, despite the immense diversity the cultivation of fish and shellfish in aquaculture systems is now dominated by B35 species that together account for 90% of total global production. Four species alone (grass carp, silver carp, Indian carps, cupped oysters) account for B30% of global aquaculture output by volume.

Aquaculture typically involves the enclosure of a species in a secure system under conditions in which it can thrive. Interventions in the life cycle range from exclusion of predators and control of competitors (extensive aquaculture) to enhancement of food supply (semi-intensive) or even the provision of all nutritional requirements (intensive). Intensification of production also implies increasing the number of individuals per unit area, which decreases the local demand for land/sea space but instead requires greater use and management of inputs, management of waste products and a greater reliance on technology and fossil energy. Aquaculture is an economic activity that uses and transforms natural aquatic resources into commodities valued by society and in so doing it may impact on biodiversity, essentially due to the consumption of resources, the transformation process (aquaculture), and the production of wastes. Looking at the diversity of farming systems it is easy to appreciate that the biophysical impacts of aquaculture activities, that is, magnitude and spatial scale, vary enormously. Technical and economic inputs, such as construction materials and energy, in many traditional aquaculture systems form only a small part of the inputs needed. The main and critical inputs are instead natural resources and to some extent also labor. Together with ecosystem services these ultimately determine the limits for the local and global expansion of aquaculture. The magnitude and type of resource use and impacts of aquaculture are, however, very much dependent on species cultured, farming system, intensity of farming methods, and management.

Group	System	Method of Culture
Plants Eucheuma, Kappaphycus, Gracilaria, Gelidium, Caulerpa	Stakes, rafts, long-lines, beds	Extensive
Bivalve mollusks Oysters (Crassostrea, spp.), mussels (Mytilus spp.), Cockles, Abalone	Rafts, long-lines, stakes, beds, tanks	Extensive
Shrimps and prawns Shrimps (Penaeus spp., Litopenaeus vannamei	Ponds	Extensive, semi-intensive, intensive
Marine, brackishwater fish Milkfish (Chanos chanos), yellowtail (Seriola spp.), groupers (Epinephelus spp.), mullets (Mugil spp., Liza spp.), cobia (Rachycentron canadum	Ponds, cages	Semi-intensive, intensive

Freshwater/diadromous fish Chinese carps, Indian carps, tilapia, Atlantic salmon (Salmo salar), trout, catfish (Pangasius)	Ponds, cages, tanks	Extensive, semi-intensive, intensive

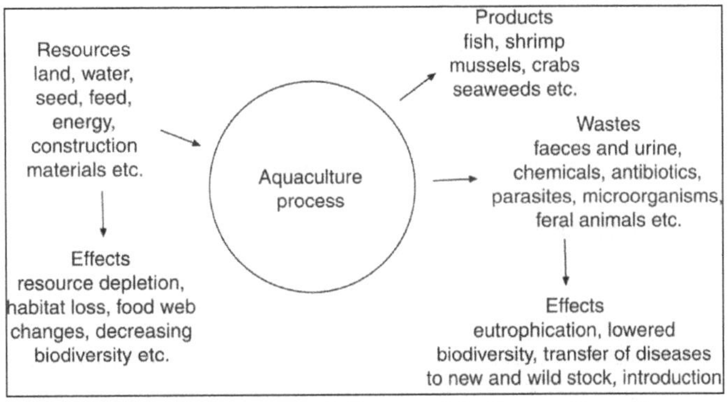

Figure: Diagram illustrating the principal direct and indirect effects from aquaculture on biodiversity, through the use of resources and the generation of wastes.

It should be noted that additional impacts associated with specific culture systems exist, and that as impacts resulting from management are not included. Note that other impacts being specific to system.

Production practices and their impacts on aquatic ecosystems vary widely across species. Mollusks such as scallops and mussels are generally farmed along subtidal or intertidal coastlines where wild-caught or hatchery-reared seed are grown in bags set on the sea bottom or on stakes and suspended ropes. The animals rely entirely on prevailing supplies of plankton and organic particles for food. Even though no feed is added, large densities of filter-feeding organisms may cause reversible local organic accumulation, resulting in negative effects on benthic diversity, and possibly also, if farmed in large quantities, impact on pelagic plankton communities. A range of systems – ponds, tanks, or cages – are used to farm finfish. The majority of carp and other freshwater species farmed in the tropics and subtropics are herbivores/omnivores and are grown in ponds fertilized by supplemental feeds. In contrast, most diadromous and marine finfish, including both tropical and temperate species, are farmed intensively in floating net cages and are to a large extent reliant on nutritionally complete fishmeal and fish oil-based diets. For some marine fish species farmed mainly in Southeast Asia, including also freshwater species like Pangasius catfish (e.g., in Vietnam) large amounts of fish, so called trash fish, are being used directly as feed.

Penaeid shrimps, dominating crustacean farming, are reared in semi-intensive or intensive coastal pond systems. The shrimps depend mainly on formulated pellet feeds, aeration to replenish dissolved oxygen, and pumped seawater to dilute pollutants and flush out harmful metabolites. Shrimp postlarvae are either derived from captured wild spawners/brood stock or directly collected from the sea – something that is true also for the freshwater shrimp Macrobrachium.

The aquaculture process in itself may affect biodiversity as a result of disturbance through increased road and boat traffic. High densities of farmed fish and food often attract predators and scavengers such as wild fish, gulls, and seals. These can come into conflict with farmers and may be killed, either accidentally (entanglement in nets) or deliberately (shooting and trapping), or if they become established they may displace sensitive local species.

Feed Resources

Extensive or semi-intensive aquaculture, for example, pond farmed carps and filter-feeding bivalves, depends either on natural production or agricultural wastes and some generally locally made feed. Even if filter feeders like mussels do not depend on addition of feeds dense farming in semi-enclosed coastal areas may impact on the food web. In Rio Arosa, Spain, for example, overgrazing of the phytoplankton population by filter-feeding mussels resulted in zooplankton starvation and the subsequent collapse of the sardine fishery.

Nutrient-rich materials added to stimulate the growth of algae and other food items, and on-farm feeds, based largely on cheap, locally available agricultural by-products, augmented by household scraps and perhaps small amounts of fishmeal, are used to supplement the food in extensive ponds. However, in the intensive production systems that predominate in temperate aquaculture, and are increasing in the tropics, the farmed animals are to varying degrees reliant on nutritionally complete commercial feeds containing fishmeal and fish oil. Marine fish species and shrimps have high demand for fish resources in the feed but also freshwater fish species like carp, tilapia, and catfish, which being herbivores or omnivores, are also increasingly being farmed using formulated feeds containing various percentage of fishmeal and fish oil. In 2010, up to two-thirds of the world's farmed finfish and crustaceans were dependent on commercial pelleted diets.

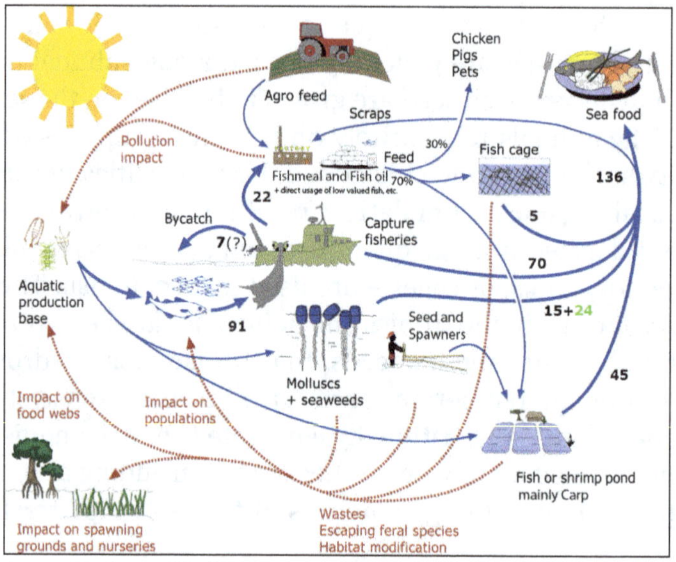

Figure: Ecological links between aquaculture and capture fisheries.

Diets for salmonids, sea bass, sea bream, and other carnivores are largely composed of fishmeal and fish oil. Although it may be possible to replace much of the fishmeal used in intensive fish diets with proteins of plant origin (e.g., oilseeds), requirements for essential amino acids, especially cystine and methionine, is still to a large extent being met by fishmeal. It remains to be seen whether commercial plant protein-based diets can be developed in an industry in which the product is competing with many others for customer attention and in which profit margins are increasingly being squeezed. Depending on the source and inclusion rate, oilseed meals can compromise palatability, growth, and profitability. Any decrease in palatability or diet digestibility may aggravate waste loadings to the environment. The issue of fish oils is even more pressing than that of fishmeal. Aquatic carnivores are poor at using carbohydrate to supply energy requirements, relying instead on protein and lipid. The substitution of fish oils with vegetable oils in freshwater carnivorous or omnivorous fish diets is more straightforward than for marine and diadromous carnivorous species, such as Atlantic salmon, which require n-3 highly unsaturated fatty acids. Progress has been made with respect to alternatives sources like rapeseed and linseed as well as synthesis of n-3 fatty acids, which can supplement various feeds while still resulting in acceptable high growth and quality.

Thick blue lines refer to main flows from aquatic production base through fisheries and aquaculture to human consumption of seafood. Numbers refer to 2012 data and are in units of megatons (million metric tons) of fish, shellfish and seaweeds (green). Thin blue lines refer to other inputs needed for production. Hatched red lines indicate negative feedbacks. Flows to, for example, pond aquaculture from cage aquaculture also exist as used feed fish generates surplus of fishmeal.

Currently, approximately 25% of the total harvest from capture fisheries is destined for nonfood use, of which most is used to produce fishmeal and fish oil for aquaculture. Fish species being used for reduction to fishmeal includes species such as Peruvian and Japanese anchovy, Blue whiting, Atlantic herring, chub mackerel, Chilean jack mackerel, capelin, European pilchard. Many fishes exploited for feed are fast growing opportunistic species and therefore can sustain a heavy fishing pressure. However, production of some of these species is also constrained by climatic variability associated with El Nino–Southern Oscillation events and impacts on marine biodiversity from intensive fishing on pelagic forage fish species, including reduction of available food supplies for marine predators and valuable species consumed by humans have not been studied in a systematic way. In Europe, the crash of North Sea capelin and herring stocks has been attributed to over-fishing and this may have caused depletion of other wild fish stocks (e.g., cod) and the starvation of seals and seabird chicks. Declining capelin populations in the western Gulf of Alaska are implicated in the decrease of harbor seal and sea lions in the early 1980s. A strong interaction between anchoveta and seabird and mammal populations has also been well documented for the Peruvian upwelling system. Recent focus on fishing for krill near Antarctic for aquaculture feeds may open up new resources that can be extracted, but as these systems still are not fully understood and especially how they will be impacted from climate change, any extraction should be carefully evaluated.

Low value fish, that is, so-called “trash fish” (from rivers, lakes, and the sea) has increasingly been used directly as feed in aquaculture, especially in Asian countries, with implication for both biodiversity and food security. Despite significant progress being made on reducing inclusion of fishmeal and fish oil in feeds, and finding alternative feed ingredients (e.g., plants and microorganisms), some intensive and semi-intensive aquaculture systems use more fish protein to feed the farmed species than is ultimately harvested. Also, the increase in global aquaculture production has resulted in a net increase in total demand for fish resources in aquaculture and it is estimated that 68% and 98% of global fishmeal and fish oil, respectively, were utilized by the aquaculture sector in 2006. Alternatives to fishmeal and fish oil include soy meal, soy oil, livestock co-products, vegetable oil, single cell, and other locally available resources (e.g., snails in Bangladesh). Other commonly used feed ingredients for many freshwater fish species include rice and wheat bran, maize gluten meal, and cassava meal.

In 2006, 74.2% of total aquaculture production was of species feeding low in the aquatic food chain, including aquatic plants, filter-feeding mollusks, and herbivorous and omnivorous finfish species. Farming of fish species such as carps, tilapia, and catfish dominates production and even though this production is increasingly based on fish resources, it results in a net production of fish. This will, however, change if farming methods continue to intensify and increasingly utilize higher quality fishmeal-based feeds or feeding with low-valued fish. For other fish species, for example, marine and some diadromous species (e.g., salmon), production requires more fish as feed than is ultimately produced. For example, approximately 3.4 kg of wild fish is used to produce 1 kg of farmed salmon. The culturing of such species thus leads to a net loss in fish protein and fish oil.

Human consumption of seafood was 136 mmt in 2012, of which 70 mmt of fish, crustaceans, and mollusks come from capture fisheries, whereas 67 mmt are from aquaculture. Total capture fisheries was 91 mmt but to this should approximately 7 mmt of discarded catches be added and of this 22 mmt are being used for fishmeal production or as direct feed. An increasing volume of processing wastes from aquaculture and fisheries are converted into fishmeal. Even though farming efficiency improves and substitutes for fish in feeds are being developed, an increasing proportion of fish resources will probably be used for aquaculture feeds as supplies are unlikely to expand and as aquaculture production continues to grow (and demand for higher trophic level species increasing) and production methods of pond fishes in major producer countries such as China intensify.

Land, Water and Energy

Land is needed to build fish or shrimp ponds or establish tank-based operations, whereas fish cages, pens, and mussel and seaweed farm operations occupy areas of lakes, rivers, and the sea. Fish ponds are usually sited in agricultural land and this arguably contributes positively to the floral and faunal diversity of agriculture landscapes. However, unproductive, boggy areas of agricultural land have often been used, and since such boundary ecosystems, or ecotones, may serve as reserves for species in areas otherwise surrounded by monocultures of crops, this may in fact reduce biodiversity. Land is becoming scarcer and the increasing competition with other users, for example, agriculture and urban development, puts pressure on aquaculture to minimize appropriation of land. This has promoted the intensification of farming systems with high stocking densities, resulting in higher dependence on external concentrated inputs, for example, feed, energy, and chemicals. On the positive side, less land is needed per metric ton of fish production and the resultant more concentrated wastes are also more amenable to treatment. However, if treatment is not done the environmental impacts may be severe.

Large areas of tropical coastal wetlands and mangroves have historically been converted to fish and shrimp ponds, resulting in impoverished biodiversity and recruitment to fisheries, with consequences for local communities and regional economies. Thus, since the 1400s, hundreds of thousands of hectares of mangroves have been transformed into milkfish ponds in Indonesia and the Philippines. In recent decades, shrimp farming has been responsible for a significant share of the conversion of coastal and supratidal areas, for example, 102,000 ha of mangrove forests in Vietnam during 1983–87 and 65,000 ha in Thailand during 1961–93. When the full range of ecological effects associated with mangrove habitat loss are accounted for, the net production in fish and shrimp aquaculture may be negative. Coastal ecosystems, such as mangroves, sea grass beds, and coral reefs, provide habitats and nursery areas for many fish and invertebrate species caught in coastal and offshore fisheries. A positive relationship between commercial fish/ shrimp landings and mangrove area has also been documented throughout the tropics. To identify and value total commercial and subsistence fisheries catch supported by mangroves, economic analyses must take into account:

- The large number of resident and transient species that utilize mangroves as habitat.
- The direct and indirect subsidies of shrimp trawlers and mangroves, respectively, to total fisheries catch.
- The aquaculture industry's dependence on inputs like seed, brood stock, and feed.

By acknowledging these support functions, the potential life-support value of mangroves to fisheries is in the order of 1–10 mmt of fish and shellfish per hectare per year (first sale value E$1000–10,000 US in developing countries). In addition, mangroves also harbor a wide array of non-marketed fish, crustacean, and mollusk species, whose subsistence harvest constitutes an important protein source for coastal communities. Different types of integrated mangrove–shrimp production systems exist in, i.e., many Asian countries. Shrimps are here stocked in ponds with mangroves planted either within the pond or on bunds and/or platforms in and around ponds. These systems are aimed to increase sustainability within the wider coastal ecosystem. Moreover, mangroves are closely linked to habitat conditions and associated biodiversity of coral reefs and sea grass beds through the biophysical interactions in the coastal seascape. Almost one-third of the world's marine fish species are associated with coral reefs, and fish catches from reefs contribute to approximately 10% of human fish consumption at a global level and much higher in developing countries.

Aquaculture requires large amounts of water to physically support the farmed animals, replenish oxygen, and remove wastes. The impacts of aquaculture on the quantity and quality of water resources have direct impacts on associated aquatic biodiversity. Large amounts of water pass through cages and pens (in the sea or lakes) but there is no net removal from the system. It is important to distinguish between water consumption and water withdrawal, the former implying that water diverted from streams, rivers or aquifers lost through evaporation or seepage, while in the latter water is returned to the environment to be reused or restored. Consumptive water use in aquaculture has mainly impacted on the reduction in downstream freshwater flows and groundwater resources.

In land-based systems, aquaculture not only borrows water, returning it in a more degraded form, but also consumes it or accelerates its loss from surface to groundwater or the atmosphere. Thus, by creating ponds, especially in areas of poor (sandy/loam) soils or high temperatures, evaporation and seepage are increased and 1–3% of the fish pond volume may be lost in this way each day (up to 45 m3 kg1 produced in ponds). Such losses may be particularly significant in arid or semiarid areas of the world, such as Israel, where fish pond design and man-agement practices have had to be changed in order to reduce surface water losses. Conversely, the incorporation of a fish pond into small rural farms has been shown to improve water conservation (i.e., creating a water reservoir). In addition, recent analysis shows that indirect freshwater consumption in aquaculture can be significantly higher than direct consumption. This is because formulated feeds indirectly consume large quantities of water through crop and livestock production. About 1.2 m3 of water is needed to produce 1 kg of grain. Fish and crustaceans require less grain compared to other terrestrial animal production, and even if water consumption through feed is large total freshwater consumption by aquaculture is small compared to, for example, agriculture. Total freshwater use depends on system and practice but it has been estimated that the 8,750,000 has freshwater and 2,333,000 has brackish water ponds for aquaculture consume approximately 429 km3 year1 (16.9 m3 kg1), which represents only 3.6% of flowing water globally.

Intensification of the aquaculture sector has led to an increasing dependence on external energy inputs throughout the production chain. Feed is commonly the main energy demanding source for fed aquaculture systems, while pumping, water purification and aeration may contribute significantly to "closed" systems. Transportation, in turn, often accounts for a surprisingly small fraction of energy inputs (less than 1% for Norwegian salmon) while holding aquaculture products at warehouses can be an energy consumption hotspot for live organisms such as mussels and abalone. Cumulative energy demand has been shown to provide a good indicator of both carbon footprint and ecological footprint while the true environmental consequences related to energy production depend largely on the energy carrier and country of production.

Wild Capture of Larvae and Spawners

The farming of shrimp and fish depends on larvae collected from the wild or reared in hatcheries from eggs of wild or farmed brood stock or spawners. Although most of the aquaculture industry today relies on hatchery-produced fry and fingerlings derived from parents bred in captivity, some tropical marine fish and shrimp culture, as well as many freshwater fish, still depends on capture of wild brood stock or juveniles (i.e., India, Bangladesh, The Philippines, Indonesia, etc.). Besides adverse effects on wild stocks of the target species, large bycatch of other larvae are killed in the process, representing losses to capture fisheries and biodiversity. The quantities of bycatch associated with such wild catches are directly proportional to the natural abundance of the target species for culture. For example, milkfish Chanos, chanos constitute only 15% of total fry in inshore collections by seine net. The annual utilization of approximately 1.7 billion wild fry for stocking in Philippine milkfish ponds in the end of the 1990s corresponded to a loss of nearly 10 billion fry of other fish species. Many shrimp and prawn ponds in India and Bangladesh are still to a large extent dependent on stocking with wild-caught seed of Penaeus monodon and Macrobrachium rosenbergeii. The peak period for this fishery reached one billion P. monodon seed collected annually in southeast Bangladesh. Up to 900 fish and other shrimp fry species are discarded for every targeted shrimp being collected from estuarine and coastal waters in Bangladesh. This is sufficiently large to potentially cause major impacts on biodiversity and capture fisheries production, but to date remains unstudied.

The development of hatcheries for cultured shrimp and marine fish species (e.g., milkfish) has reduced dependence on wild seed. However, it has also increased the demand for wild-caught mature (spawners) and immature (brood stock) shrimp adults. For example, because the species is rare, wild collection of P. monodon brood stock and spawners can lead to large amounts of bycatch, with consequences for fisheries and biodiversity.

Impacts of Wastes, Chemicals, Diseases and Feral Animals

The term "wastes" in the current context is used to mean not only food, fecal and urinary products, and chemicals, but also microorganisms, parasites, and feral animals that may be introduced and subsequently thrive in aquaculture environments which may subsequently be released to the wider environment. The release of uneaten food and fecal and urinary wastes may lead to eutrophication and oxygen depletion, the magnitude of the impact depending on the type and size of operation and the nature of the site, ecosystem characteristics and assimilative capacity. Local effects from nutrient release into marine ecosystems, from coastal and near-shore intensive aquaculture operations, especially shrimp pond and fish cage farming, have been well studied. For example, eutrophic conditions from uncontrolled fish pen and cage operations in Pangasinan, Philippines, included increased ammonia, nitrate, nitrite, and phosphate concentrations, and low dissolved oxygen levels, leading to a major fish kill in 2002 valued at US$ 9 million.

Sediment impacts associated with cage farming can be severe, resulting in anoxic sediments where fauna been lost. The effect on sediments is, however, mainly local and biodiversity will return a few years after production stops. However, studies documenting more large scale and long-term ecosystem changes from excessive nutrient emissions are few. Avoiding release of wastes, thereby minimizing environmental impacts, has been one of the drivers toward land-based "closed" production systems. This has, however, come at the price of higher energy dependency. However, semi-intensive, extensive, traditional, polyculture, and integrated systems generally assimilate much wastes internally resulting in fewer wastes being discharged to surrounding ecosystem.

Chemotherapeutants, including antimicrobial compounds and pesticides, are mainly used in intensive fish and shrimp cultures to control bacterial, fungal, and parasitic diseases. In shrimp farming, prohibited chemicals may be used due to lack of legislation or poor implementation of regulations. Farmers also use a range of vitamins, immunostimulants, disinfectants, and chemotherapeutants and employ chemicals for pond soil and water treatment. The impacts of many of these chemicals are largely unknown. The release of antibiotics into the environment can affect microbial biodiversity and antimicrobial drug resistance but has reduced hugely in the farmed salmon industry due to the introduction of vaccines. Controls on use are increasingly strict, especially for aquaculture products aimed at the European and North American market, but the controls target mainly the finalized products and not the aquaculture production process.

Impacts of aquaculture operations on biodiversity are strongly linked to introductions of exotic species and strains. Worldwide transfers and introductions of the few preferred shrimp culture species, including P. monodon, Litopenaeus vannamei, and Marsupenaeus japonicus, were numerous in the early decades of commercialized farming. Such introductions and transfers may lead to competition with endemic fauna, genetic introgression with local fauna, and introduction of pathogens and parasites. Tilapia has had a long history of deliberate (for aquaculture and fisheries) and accidental introductions to some 90 countries and territories. However, recent studies suggest that impacts of tilapia introductions in Asia on aquatic ecosystem structure and function have been relatively small and often outweighed by the socioeconomic benefits. Atlantic salmon have escaped from culture facilities both within the geographic range of wild Atlantic salmon as well as in Pacific waters, and are now found as far north as the Bering Sea and as far south as Chile. Increasing evidence suggests that escapes may have direct genetic impacts on wild populations through hybridization. Larger numbers of escapes also increase the likelihood of hybridization between feral farmed Atlantic salmon and wild fish in populations that are locally endangered or close to extinction. In addition to consequences for the population gene pool and fitness, there are many potential ecological impacts associated with feral fish. Feral Atlantic salmon may compete extensively with wild salmon species for food and space, disturb native spawning sites, and introduce new diseases and parasites into wild populations.

The consequence of the large-scale introduction of the shrimp Penaeus vannamei into Asia and the Pacific on biodiversity of indigenous cultured or wild shrimp populations is still uncertain as insufficient time and research have been conducted on this issue and there is therefore a need for caution. Recent field surveys in the Bangpakong Estuary in Thailand (where this species contributed 499% of total 2007 marine shrimp production) show that the species is persistent in the wild, and the presence of Taura Syndrome Virus (TSV) has been detected in seven local species. Moreover, laboratory studies show that P. vannamei can tolerate environmental conditions in the estuary and seeks food better than some of the indigenous species. The development of genetically modified species for aquaculture is currently being heavily debated, and of particular interest is the "super salmon" that has been developed by Aqua bounty and has just been approved by the US Federal Drug Administration to be farmed commercially. The animal has a single copy of a DNA sequence that includes code for a Chinook growth gene as well as regulatory sequences derived from Chinook salmon and ocean pout.

Numerous wild salmon stocks in Canada and Norway have been infested by sea lice, either through the release of juvenile farmed salmon or large densities of fish cages in coastal salmon migration routes. The mechanisms and linkages to farmed salmon in Canada, however, are still being debated.

Diseases are prevalent in many types of aquaculture, especially in highly stocked, intensive production systems, for example, Atlantic salmon, marine fish species, marine shrimp, etc. For example, the introduction of shrimp postlarvae and brood stock from areas affected by the White Spot Syndrome Virus (WSSV) and TSV resulted in the rapid spread of these pathogens throughout most of the shrimp-growing regions in Asia and Latin America, respectively. A native of Asia, where it has caused multimillion-dollar shrimp crop losses, the WSSV has also been detected in wild and cultured shrimp in Texas and South Carolina. The virus was probably introduced by release of untreated wastes from plants processing imported Asian shrimp into coastal waters and by use of imported shrimp as bait in sport fishing or as fresh food for rearing other aquatic species. Another major shrimp virus, the Infectious Hypodermal and Hematopoietic Necrosis Virus (IHHNV), was probably introduced to the Americas from Asia through the import of live P. monodon in the early 1970s. In the Philippines, IHHNV prevalence in various wild populations of P. monodon has been correlated with intensification of shrimp culture methods and with mangrove status. Higher incidence of viral infection in wild shrimp has been observed in areas with intensive shrimp farms and severely degraded mangroves.

The multibillion dollar Chilean salmon industry was during 2008 hit by Infectious Salmon Anemia (ISA) virus, resulting in millions of dead salmon and the layoff of thousands of fish farm workers. The main salmon farming areas in Chile are around Chiloe Island and this area is now being contaminated. Salmon producers are therefore reducing farm and fish densities as well as seeking to transfer their businesses further south. Whether the move will be successful or not is uncertain because this area has previously been almost free of farming operations due mainly to its remoteness and concern for the integrity of the seascape. Another factor that may compromise the successful translocation of salmon farming to the south is that the ISA virus already appears to be present in these waters.

In contrast to shrimp and salmon, comparatively few diseases have been reported for carps, tilapia, and milkfish, particularly from extensive and other low-density culture systems. The current trend toward intensification in rearing ponds and cages, however, may create stressful conditions through deterioration of water quality, excessive stocking, and polluted water inflow that predispose the fish to disease. The farming of Pangasius catfish in Vietnam has increased rapidly, reaching more than 1 mmt in 2008, and farming is characterized by holding high densities of fish in the ponds, resulting in increased risks for diseases and heavy usage of antibiotics.

Resource use and Carrying Capacity – Energy Analysis, Ecological Footprint and Life-Cycle Assessment

To reduce the risk of resource constraints and impacts on biodiversity, a shift to aquaculture production systems that use less valuable resources and emit wastes that do not exceed the assimilative capacity of the environment must occur. There is a need to recognize and manage nature's life support and its provision of ecosystem services, on which economic development and human welfare depends. Besides traditional EIA one way of identifying the broader demands for natural resources and ecosystem services of aquaculture has been to estimate the ecosystem area – the ecological footprint – functionally required to support the activity. When problems beset aquaculture operations, solutions focus on the pond or cage unit, and the fact that the farm is part of a much larger ecosystem with which it interacts is generally not considered. Surrounding ecosystems provide the feed, seed, clean water, and other essential resources and services, including waste assimilation. This unvalued work of nature sets the limits to culture levels without compromising biodiversity or causing pollution or disease problems. The footprint concept has proven to be very useful in illuminating the non-priced and often unperceived work of nature that forms the basis for economic activities such as aquaculture. It is, however, not a detailed tool for measuring environmental capacity or sustainability in aquaculture, but an excellent tool for communicating human dependence on life-support systems.

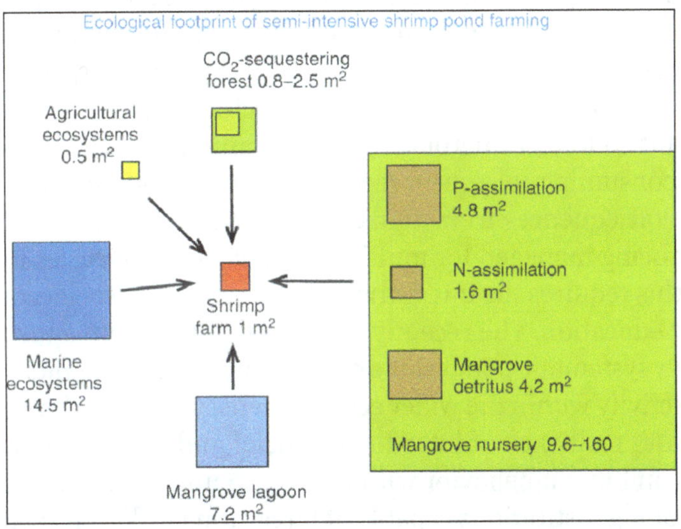

Figure: Ecosystem support areas required to sustain a semi-intensive shrimp farm in a coastal mangrove area of Columbia/square meters of support area needed per square meter of pond area.

An illustration of how the footprint concept can be used for aquaculture is provided by Larsson et al. They estimated the spatial ecosystem support, or footprint, for a semi-intensive shrimp farm in a coastal mangrove area in Colombia. Support included food inputs, nursery areas, clean water, as well as waste processing and the support area was 35–190 times the surface area of the farm. The details of the footprint include: mangrove nursery area required to produce the shrimp seed for stocking – up to 160 times the pond area; if located close to the farm, the same mangrove area could also supply natural food inputs (4.2 m^2 m^2 shrimp pond area) and absorb polluting nutrients (2–22 m^2 m^2 pond area) in the farm effluents; feed pellets, a major input to a shrimp farm, needed a marine area of 14.5 m^2 to produce the fish, and an additional agricultural area of 0.5 m^2 for the agriculture ingredients used in feed pellets; 7.2 m^2 was needed to provide clean lagoon water to the ponds; and 0.8–2.5 m^2 of forest area per square meter shrimp pond area was needed to sequester the CO_2 associated with fossil fuel consumption at the farm.

Footprint size changes with farming intensity that, is, a higher stocking density requires more food inputs and produces more wastes. However, less intensive production operations stock lower densities of species and thus require larger areas to produce the same biomass of produce. Pressure on local ecosystems can be reduced to some extent by importing some inputs (e.g., feeds) from other areas and by investing in shrimp hatcheries. Although producing seed in hatcheries considerably reduces shrimp and fish larvae bycatch that would otherwise be recruited populations exploited by fisheries, it also increases demand for wild-caught spawners and brood stock. Other ecosystem services, however, such as clean water supply and waste assimilation, must be located close to the farming area. Up to a certain level of farming intensity this may pose little problem, but the whole operation may collapse when the dynamic carrying/assimilating capacity of the local environment is exceeded unless extensive and costly water treatment facilities are built. The footprint concept provides an early warning device when the level of carrying capacity is being approached. Integrated farming technologies that re-circulate resources and wastes within the farm may be one way of reducing the footprint.

Aquaculture has directly contributed to the loss of important ecosystem functions (and biodiversity) through land and seascape transformation, and also more indirectly through, for example, pollution. However, aquaculture has also enhanced provisioning services, both in the agriculture landscapes and in the seascape, thus leading to improved welfare through livelihood diversification. Aquaculture can be a viable substitute for some sectors of today's terrestrial animal production (ie, as an important source of micronutrients and animal-based protein and lipids), proven to be highly resource consuming (e.g., beef feedlot systems). The question is how to balance the negative and positive consequences from aquaculture development. The landscape and seascape are today increasingly being managed for multiple functions and services in addition to provision of food and fiber, and this requires the integration of ecological and socioeconomic research, policy innovation, and public education. The recently developed "Ecosystem approach to aquaculture" is an example of a broader systematic perspective on aquaculture development. This is a "strategy for the integration of the activity within the wider ecosystem in such a way that it promotes sustainable development, equity, and resilience of inter-linked social and ecological systems," and could potentially force changes in human behavior with respect to understanding ecosystems functioning and the need for developing institutions capable of integrating different sectors at multiple scales.

Life cycle analysis (LCA) is a standardized eco-efficiency measurement framework that is increasingly being used to build inventories and estimate one or several environmental impacts in aquaculture.

LCA identifies system boundaries at the broadest scale, both temporal and spatial, and is a much more quantitative tool than Ecological Footprints. Thus, the width of environmental impacts that can be measured allows for more detailed estimates of the environmental interactions. Climate change, habitat change, pollution and overexploitation are impact categories commonly entailed within the LCA framework, addressing all major drivers, except invasive species, for biodiversity loss, identified by the Millennium Ecosystem Assessment.

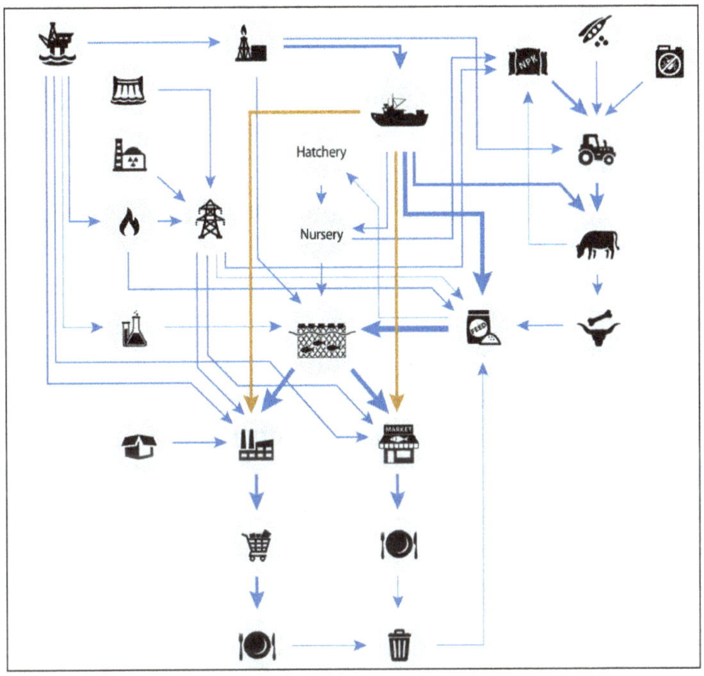

Figure: Illustration of system boundaries being identified in Life Cycle Analysis of an aquaculture production system.

Integrated Aquaculture

Integrated aquaculture may offer opportunities for the efficient usage of water and utilization of nutrients, and increased productivity and profits, providing in a single package practical and creative solutions to some problems of waste management and pollution. Thus, the resulting environmental impacts from aquaculture and various resource limitations (water, feed, energy, etc.) may find their solutions in integrated cultivation techniques. Traditional inland aquaculture, comprising diverse systems that use on-farm or locally available resources, are still common in Asia, but they are increasingly being replaced by industrial-based aquaculture technologies. Traditional polyculture/integrated systems make efficient use of inputs and generate less waste, thus adding to net food supplies locally and regionally at relatively minor environmental and social expense. Traditional pond cultures of herbivorous fish species (e.g., carps in China) have been viable for centuries and their existence is the proof of sustainable integrated farming systems. Here, raising poultry and livestock is integrated with fish farming, and the principal linkages between the systems are animal manure and other agriculture waste products. Compost is used to fertilize the pond water for proliferation of natural organisms as natural feeds for fish from juvenile to adult. However, despite its environmental advantages the increased global demand for food cannot be met by traditional extensive production systems (due to demand on space and to low productivity). Even though technological development and improved management has resulted in increased efficiency and environmental performance of some intensive single species aquaculture systems, we need to ask what information embedded in traditional integrated systems might be useful for aquaculture development. This information could then be added to recent research on integrated aquaculture systems. Integrated aquaculture is certainly not a panacea for aquaculture development, but should be looked upon as one potential approach among many others facilitating sustainable development.

The concept of ecological engineering has gained interest in an aquaculture context, with negative environmental effects being remedied by species integration for nutrient trapping or recirculation. The recent development and promotion of integrated aquaculture in coastal areas has focused on modern integrated approaches, mainly from temperate and subtropical regions (China, Canada, South Africa, the Mediterranean Sea, i.e., IMTA systems (Integrated Multi-trophic Aquaculture)) involving integration of fish, seaweeds, crustaceans, and mollusks. Such systems try to tackle the throughput characteristics of many farming systems, in which large amounts of wastes are released to the environment. Thus, in addition to output of particulate wastes, aquaculture also releases dissolved nutrients, and generally less than one-third of the nutrients added through feed will be removed through harvest in intensive fish and shrimp farming. Treatment of effluents usually involves a high degree of technology, and therefore high costs, which implies that release of untreated water is the rule rather than the exception (especially in many developing countries). Besides the improved environmental performance, the benefits from integration include economic gains resulting from co-production of different products.

Tropical integrated mariculture systems involve various types of polyculture, sequential integration, temporal integration, and different integration with mangroves. An example of the latter is the use of mangroves as bio filters to process effluents from aquaculture ponds. Results from the central Philippines indicate that 1.8–5.4 ha of mangroves are needed to assimilate nitrate wastes from 1 ha of shrimp pond. Mangroves and aquaculture are not necessarily incompatible and guides for responsible use of mangrove areas for aquaculture have been developed. However, any development of aquaculture within or nearby mangroves needs to be carefully evaluated to identify net benefits and impacts on environment, biodiversity, ecosystem services, and implications for a variety of stakeholders.

Basic Principles of Integrated Fish Farming

Integrated fish farming is based on the concept that 'there is no waste', and waste is only a misplaced resource which can become a valuable material for another product. In integrated farming, the basic principles involve the utilization of the synergetic effects of inter-related farm activities and the conservation, including the full utilization of farm wastes.

It is assumed that all the constituents of the system would benefit from such a combination. However, in most cases, the main beneficiary is the fishes which utilize the animal and agricultural wastes directly or indirectly as food. As integrated farming involves the recycling of wastes, it has been considered an economic and efficient means of environmental management.

Types of Integrated Fish Farming

Paddy-Cum-Fish Culture

Rice fields which are water-logged for 3-8 months in a year, there is always small population of fishes that gain access to such waters. This probably had given rise to the practice of deliberate stocking of fishes and harvesting.

- Paddy-field aquaculture provides additional income to the farmers.
- In areas where rice and fish form the staple food, paddy-field aquaculture makes available an essential diet for the people.
- As paddy and fish can be grown either simultaneously or alternately in the same water mass, it requires very little extra input by way of additional costs, particularly in management and labour.
- It provides off-season employment to the farmers and farm labors.
- Combination of paddy and fish farming is mutually beneficial. Fish cultivation promotes better paddy production by way of exercising an effective control on unwanted weeds, molluscs, noxious insects and their larval stages.

Fishes Suitable for Paddy-Cum-Fish Culture

All fishes are not suitable for such a type of culture as paddy-fields provide special ecological conditions such as shallow turbid water with high temperature. Fishes having the following criteria are generally selected for paddy-cum-fish culture:

- Fishes that can adapt to shallow waters necessary for paddy crops.
- Fishes that can tolerate high temperature.
- Fishes that can thrive on low dissolved oxygen, which is the characteristic of paddy-fields especially in tropical countries.
- Fishes that can tolerate fairly high turbidity.
- As the duration of culture is quite short, fishes that have high growth rate is to be selected, so that it can reach marketable size within these few months.
- Fishes that can live in confinement and do not tend to escape from the cultivated area.

Fishes that are cultured in such waters in India are Mugil sp., Mystus gulio, Haplochromis mellandi (mollusc eating fish), hates calcarifer, Mugil parsia, Puntius sp., Channa sp., prawns and shrimps. In India, limited experimental works have shown the suitability of Indian carps for such integrated farming.

Management of Paddy-Fields

To increase the utility of paddy-fields as fish ponds, the following managements are required:

- A continuous flow of water in the field, with proper inlet and outlet is to be maintained.
- The water in the field is to be maintained at a desired level.

- Proper drainage of water from the field has to be made in case of flooding.
- At the point of entry and exit of water, some control means (such as screens) have to be provided to prevent the cultivated species from escaping and stopping the entry of wild fishes into the paddy-field.
- Deep pits or other devices have to be provided as shelter to the cultivated fishes at the time of distress.

Types of Paddy-Field Aquaculture

Three major types of paddy-field aquaculture are generally practiced:

- Simultaneous or combined or synchronous farming.
- Alternate or sequential farming or paddy- fish rotation.
- Relay farming.

Simultaneous or Combined or Synchronous Farms

In such farming paddy and fish are grown together and harvesting the paddy and fish at the end of the rice-growing season. Literature reveals that a variety of techniques and stocking rates have been employed in different countries on paddy-field aquaculture. In general, fishes are stocked not earlier than 5 days after transplantation of paddy seedlings, so as to enable the seedlings to form proper root. After transplantation of paddy, stocking in case of fry should be done after 10 days, while for fingerlings after 3 weeks.

The methods of culture adopted are for production of either fingerlings or fish for consumption. Farmers generally cultivate local varieties of paddy which takes up to 6 months to be harvested, so that more than one crop per year can be grown. Generally organic manures are preferred over the use of fertilizers in the preparation of the paddy-fields. The advantages of simultaneous cultivation are:

- There is no additional cost for fish production.
- Fishes contribute to the enhancement of paddy production by destroying weeds, causing tillering and mineral enrichment by their digging activity, and for the fertilisation of soil by their excrement and also by the unutilized artificial feed.
- Due to fish cultivation paddy production gets increased by 5 to 15%.

The disadvantages are:

- The water depth being shallow hinders better fish raising.
- For fear of harmful effect on fishes, herbicides and insecticides uses in paddy-field is generally prevented which ultimately limits paddy production.

- Due to abrupt changes in temperature and dissolved oxygen, inadequate space and presence of piscivore birds, there is great loss of fishes which may be about 20-60%.

Progressive agricultural areas have slowly discarded paddy-fish cultivation in favour of the rotation of paddy and fish crops.

Alternate or Sequential Farming or Paddy-Fish Rotation

It is the simplest form, where flooded paddy-fields after harvest are used to raise one or more crops of fish or shrimp. In rotation system the paddy-field has to be prepared for raising fishes after the harvest of rice. To maintain the required depth of water the bunds surrounding the field have to be raised. After harvesting the rice, the stubbles are not removed.

These submerged stubbles provide the substrate for the development of fish food organisms. These stubbles undergo decomposition, thereby fertilizing the water and stimulating higher productivity. After fish harvesting, the residues remain in the soil which in turn serves as fertilizer for the paddy crop.

Shrimp Production in Paddy-Fields

Shrimp production in paddy-fields is generally practiced in areas where only one crop (July to September) of a salt-resistant variety of paddy is grown. Following the harvesting of rice, the bundhs are strengthened and sluice gates are installed to control the water supply.

At high tides, the fields are filled with tidal water, containing a large number of shrimp larvae. These larvae gain access to the fields, where they find food and shelter. To attract the larval shrimps, lamps are hung above the inlets. By this process, natural stocking of shrimp larvae is done at every high tide for two to three months. To prevent the escape of juvenile shrimps at low tides, conical bag-nets are attached to the sluice gates.

Shrimp harvesting from these fields starts in December, as by this time the early stock of shrimps will have attained marketable size. Regular harvesting is done which results in thinning of the stock, leading to better growth rate and harvesting of larger shrimps. The total yield of shrimp is around 780-2100 kg per hectare.

Advantages of Paddy-Fish Rotation

The paddy- fish rotation system has the following advantages:

- There is no water depth limitation either to paddy or fish cultivation. Shallow water depth is maintained till the harvesting of paddy, after which the water depth is raised for the culture of fishes.

- Through this adequate water level management, suitable water temperature and dissolved oxygen content can be maintained.

- After harvesting of paddy, the submerged stubbles decompose and fertilise the water. This leads to the development of fish food organisms and ultimately stimulates higher productivity.

- The interval between paddy harvesting and fish stocking is sufficient enough to allow degradation of pesticides.
- Insect pest infestation gets reduced as their life-cycles are disrupted.
- Either monoculture or polyculture can be practiced.

The disadvantages are: Due to presence of piscivorous birds like herons, cormorants, etc., considerable loss of fish (20-60%) takes place.

Relay Farming

This system is much complicated and it ensures prolonged period of fish culture. It involves transferring the stock of fishes to specially prepared ditches, channels or pools at the time of paddy harvest and re-stocking them in the field for a further growing period. This system provides better growth rate than what is possible in such a short duration of one paddy crop.

For such cultivation the farm should have additional holding or rearing facilities for the period between paddy harvesting and planting of new fish seedlings. The carps or shrimps can be grown to the required size through adequate feeding. This culture involves higher inputs including labour.

Duck-Cum-Fish Farming

Although duck and fish farming have been in practice in Eastern Europe and China and their compatible nature being recognised since long, the interaction and benefits of the association have been understood only a few years back.

During the last decade suitable methods have been developed in various countries for raising ducks in fish ponds. Undoubtedly, such a combined culture is highly profitable as it greatly increases the production of protein in terms of fish and duck per unit area.

The combination of duck and fish farming is seen presently as a means of reducing the cost of feed for ducks, at one hand, and on the other, the excreta of ducks acts in an inexpensive way of fertilizing ponds, which results in production of fish food organisms.

Thus, ducks can be said to be 'living manuring machines'. The duck droppings contain 25 per cent organic and 20 per cent inorganic substances along with a number of elements such as carbon, nitrogen, phosphorus, potassium, calcium, etc.

Moreover, ducks by agitating the shore areas of the pond help to release nutrients. Besides this, ducks feed on a variety of organisms such as weeds, snails, unwanted harmful insects and their larvae (some being vectors of fish pathogenic organisms), tadpoles, frogs and water-borne disease-causing organisms infecting man, whose eradication is one of the important aspects of farm management.

Ducks Culture and their Raising

Ponds provide a clean and healthy environment for ducks. Special strains of ducks have been developed that are suited for pond raising. Such suitable strains when used, approximately 50-60 per cent of their droppings when falls into the pond, is sufficient to fertilize the water.

Reliable and timely supply of required strains of good quality ducklings are of great significance as it promotes successful farming. For supply of ducklings, small scale farmers have to depend on outside sources, while larger farms have their own breeding centers which are much convenient and profitable.

Raising of Ducks is done by One of the Two Ways:

- Extensive Raising: In extensive raising ducks are stocked at approximately 150-500 per hectare of pond surface. Here small amount of supplementary feed is provided and the number of ducks is limited due to the food they can find in the pond water. As the number of ducks are limited, they contribute little amount of manure to the pond and its effect on fish yield is also limited. Such a type of method is usually employed in Europe.
- Intensive Raising: Here ducks are stocked at a much higher density, approximately 1000-2500 per hectare of pond surface and is usually employed in Africa. The ducks are fed at the same rates as on land. As ducks are held at a much higher density per unit area, higher amounts of manure are thus loaded into the fish pond which subsequently results in higher yields.

Duck Keeping

There are two basic ways of keeping ducks in fish ponds:

Free-Range System

Here the ducks are allowed to have free access to the whole pond. As they are allowed to swim freely around the whole pond surface, a good proportion of their droppings fall directly into the pond water and are distributed more or less uniformly.

The ducks are able to forage around the whole pond. Adjacent to the ponds, small duck houses are built with facilities for providing them with supplementary feed. In this system, as considerable energy is spent by the ducks in swimming activity, it is believed to affect the feed conversion ratio and growth rate of the ducks.

Figure: A fish pond where the ducks have free access to whole water body.

Confinement in Enclosures

Here, part of the pond area and the adjacent land area are enclosed with wire fences. About one-fourth of the enclosure will be on the land and the rest in water. Within the wire fence suitable feeding and resting areas are made.

Some of the droppings of the duck fall directly into the pond water, while the rest that drops on land is washed down into the pond. Under this system fish production is reported to be almost equal to the free-range system. Keeping ducks in enclosures is preferred by most farmers, who use special strains of ducks for better growth.

Raising of Fish

For duck-fish farming the most suitable pond is the barrage type of ponds. This is made by damming shallow valleys, so that the ducks can lie on the natural slopes. There is every likely-hood that ducks may damage the earthen dykes while foraging. The problem can be solved with proper maintenance.

Commonly the fish species cultured are herbivores and omnivores. The common carp was traditionally the main species but subsequently other species of Chinese carps are used to make full use of the food resources. Other important species used are male or hybrid tilapia, grey mullet, catla, eel, Asian catfish and sea perch. Fairly high stocking rates are adopted and supplementary food is also provided.

Advantages and Disadvantages of Fish-Cum-Duck Culture

Advantages

- There is practically less additional cost for fish culture, as the excreta of the ducks fertilize the pond water.
- Ducks by agitating the shore areas of the ponds help to release nutrients.
- The cost of feed for the ducks gets reduced.
- The ducks help to eradicate vectors of fish pathogenic organisms and water borne diseases — causing organisms infecting man, as they feed on a variety of organisms such as weeds, snails, unwanted harmful insects and their larvae, tadpole, frog, etc.

Disadvantages

- The fingerlings released should be of more than 10 cm size, otherwise the ducks may feed on the fingerlings.
- While foraging for food, ducks sometimes damages the earthen dykes. This problem can be solved with proper maintenance.

Fish-Cum-Poultry Farmings

Integrated fish farming with poultry is generally cultured as the poultry manure is a very efficient fertilizer for fish ponds. The poultry droppings comprise 2% nitrogen, 1.25% phosphoric acid and 0.75% potash. The low feeding cost per individual bird makes poultry farming along with fish, a common investment for poor farmers.

Poultry Raising

Both 'broilers' and 'layers' variety of chicken can be raised for fish-poultry farming. One day old chicks are raised up to the pullet stage after which they are put in layer cages. Fish culture with both intensive and extensive poultry productions have been integrated successfully. The most intensive type of poultry production is the battery type of housing, which is installed by the side of the pond. The floor of the house is cemented and is set up at a slope so that the eggs may roll forward.

For layers, the floor area required is about 30 cm^2 while for broilers, 15-20 cm^2. The usual floor space allotted for each bird is 20 x 30 x 40 cm. The birds are confined to cages which are made up of standard, stout, galvanized wire. The cages are kept on trays for collection of droppings. For manuring one hectare pond water, the droppings of about 250 layers and four batches of 200 broilers each are adequate in a year's time.

Fish Raising

For fish raising the ponds are stocked with fingerlings of catla, silver carp, common carp, murrels, tilapia, giant freshwater prawns, etc. The stocking density of fishes is related with that of poultry and also with the period of culture.

In one hectare pond area, when stocked with 5000 giant freshwater prawns (Macrobrachium rosenbergeii) and 1500 silver carp, and cultured for a period of four months, one can harvest 600 kg of prawns and an equal amount of fish, along with 250 culled birds.

For culturing over a period of one year, ponds may be stocked with fingerlings of catla, common carp, silver carp and grass carp at a density of 5000-6000 fingerlings per hectare. At the end of twelve months of fish-cum-poultry culture, fish yield of over 3900 kg per hectare can be normally obtained along with 42,000 eggs and 200 culled birds.

Advantages and Disadvantages of Fish-cum-poultry Culture

Advantages

- Chicken manure is a very efficient fertilizer, so no chemical fertilizer is needed for fertilizing the pond water. This cuts down the expenditure of rearing fishes.
- No supplementary fish feed is required.
- The purchase and feeding cost per bird is low.
- Chicks are readily available and their productivity can be improved with simple and cheap management.

Disadvantages

- Chicks should be examined from time to time and diseased one should be isolated, otherwise they will destroy the entire stock.

- Sufficient time should be given from one stocking of chicks to the next for renovation of the house and disinfectioning it.

Fish-cum-pig Culture

In integrated fish farming with pig, the 'pig dung' is useful for conditioning the soil and providing the necessary nutrients required for fertilizing the pond water. Fish-cum-pig culture is practiced at large in China where pigs are considered as "costless fertilizer factories".

Pig dung contains about 70 percent digestible feed for fishes. The feed while passing down the pig's alimentary canal gets mixed up with enzymes which continue to act even after defecation. Such undigested solids serve as direct food source for tilapia and common carp. In tropical fish ponds, weeds are a major problem in fish culture. Such vegetation's are considered as valuable food resource for pigs. Thus, Pigs Aptly Play a Role in Biological Control of Weeds

Pig Rearing

In fish-cum-pig culture the embankments of fish ponds are made wider (over 10 m in China) to facilitate the building of pig sites and also for growing vegetables, fruit trees or other crops. In the slopes, grasses can also be grown which is used as fodder for grass carp and for other farm animals.

Various aquatic plants, such as azolla, duck weed, Pistia, Wolffia, Lemna, and water hyacinth (Eichhornia) are grown in feeder channels and irrigation ditches associated with the pond farms. These and the foliage of other terrestrial plants (vegetables, rice, corn, etc.) are used for feeding the pigs.

In the water area of the pond, about 10 tons of aquatic plants can be produced, which are sufficient to feed 10 pigs. These plant materials are generally mixed with bananas, coconut meal, rice bran, soybean wastes, groundnut cakes, fish meal, etc.

Pig sites or pen or sty are generally built on nearby land or on the pond embankment. The pen enclosure is built not only for pig raising but also special consideration is given to the needs for breeding, nursing and fattening activities. Pig pen generally have a system of channels for transferring the organic matter into the pond water.

Alternatively, the sty or pen may be constructed above the pond water. The structure is made of wood and provided with a lattice type of floor which permits the excreta and uneaten food to fall directly into the pond water.

Modern practices are to avoid direct washing of the wastes into the pond. The urine and dung of pigs are first allowed to the oxidation tanks (digestion chambers) where sedimentation and fermentation of the manure take place. The supernatant liquid, at regular intervals, are then discharged into the fish ponds.

The sludge that remains is utilized as fertilizers in agriculture. Alternatively the pig manure may be kept in a heap on the pond embankment for later use. The chemical composition of pig wastes is depicted in Table below.

Source	Organic matter (%)	N (%)	P_2O_5 (%)	P (%)	K_2O (%)	K (%)
Fresh manure	15.00	0.60	0.40	0.18	0.44	0.37
Urine	2.00	0.30	0.12	0.05	1.00	0.83
Air dried manure	34.32	2.12	0.98	0.43	2.45	2.03
Litter manure	34.00	0.48	0.24	0.11	0.63	0.52

The number of pigs to be raised per hectare and the manuring rates to be applied are based on years of experience. The production of manure depends upon the age and size of the pig. A piglet produces about 3.4 kg manure a day, while a one-year-old pig gives about 12.5 kg a day. The average production of faeces and urine per pig is about 7.8-8 tons per annum. A density of 60-100 pigs has been found to be sufficient to fertilize a one hectare fish pond.

Fish Rearing

Polyculture is commonly practiced in such integrated farming due to the variety of food that becomes available in the pond. Herbivorous and omnivorous fishes are used for culture; generally common and Chinese carps and less frequently catfishes (Pangasius), Indian carps and tilapia. Due to high productivity of the ponds, fairly high rates of stocking are generally practiced — 60,000 fingerlings of different species (weighing 20-30 gm) per hectare.

Production and Duration of Culture

The duration of culture of fishes and pigs varies. Generally it is about one year, but culture for 6 months duration is also practiced. The overall economics of combined fish and pig rearing depends on the local conditions. However, the pigs are generally sold when they have attained a weight of 90-100 kg. The production of fish generally varies between 2 and 18 tons per hectare per annum.

Advantages and Disadvantages of Fish-cum-pig Culture

Advantages

- Such integrated farming increases the productivity per area and thus, the farmers income becomes doubled or more.
- Pig dung conditions the soil of a new pond and provides ready-made organic matter, containing the necessary nutrients.
- Pig dung contains about 70 per cent digestible feed for fishes. The undigested solids present in the faeces of pig serves as direct feed source for tilapia and common carp.

- Pigs aptly play a role in biological control of weeds, as weeds are considered as valuable food source for pigs.

Disadvantages

- Addition of too much pig manure may lead to increased nutrient load resulting in pollution of the water body and mortality of the fishes.
- Considerable care and management skills are required to prevent pollution. It has been found that satisfactory fish production can be obtained with much lower manuring.

Other Animals in Fish Systems of Integrated Farming

- Fish-Cum-Cattle Farming: Cattle wastes and washings from the cattle sheds are conveyed through pipes into the ponds which act as good fertilizer. Cattle wastes are generally collected in a pit for later use. In addition to fish yield, production of milk from cattle and beef adds to the economy.
- Fish-Cum-Rabbit Farming: Rabbit farming has been found to be ideal for integration with small- scale fish culture. Rabbit manure has greater value as a direct food for fish compared to other livestock wastes.
- Fish-Cum-Mulberry Farming: Mulberry plants are raised on the dikes of the fish farm and in the neighbouring fields for silkworm production. The mulberry wastes and silkworm larvae and pupae (after removal of silk) are used as feed for the fishes. It also fertilizes the pond water.

Advantages of Integrated Fish Farming

The advantages of integrated fish farming are as follows:

- Full utilization of farm wastes.
- Utilization of the cooperative effects of interrelated farm activities.
- It increases employment opportunities.
- It increases nutritional source for the farmer's family.
- It gives higher and stable farm productivity and there is less risk (biologically and economically).
- It increases the income of rural population.
- It is a means of land reclamation in certain areas.
- It is an efficient and economic means of environmental management.

Disadvantages of Integrated Fish Farming

Recently, controversy has arisen among scientists on the public health aspects of integrated farming. Speculations are ripe that integrated fish farming with pigs and poultry may be a cause of influenza pandemic. This may be, as the pigs would act as 'mixing vessels' for avian and human influenza viruses, it can create new lethal strains of viruses by mutation.

In such an act, fishes themselves do not play any role. However, there is no conclusive evidence that integrated farming can become a public health hazard. For safety measures, pig-poultry combinations in integrated fish culture should thus be avoided.

Sustainable Aquaculture

"Sustainable development is the management and conservation of the natural resource base and the orientation of technological and institutional change in such a manner as to ensure the attainment and continued satisfaction of human needs for present and future generations. Such sustainable development (in agriculture, forestry, fisheries sectors) conserves land, water, plant, and animal resources, is environmentally non-degrading, technically appropriate, economically viable, and socially acceptable."

Sustainability refers to the ability of a society, ecosystem, or any such on-going system to continue functioning into the indefinite future without being forced into decline through exhaustion or overloading of key resources on which that system depends. In general, the concept of sustainable development is simple and important, but translating it into specific standards or criteria is difficult, often subjective and misused. Although many specific sustainability criteria have been proposed there are no single universally agreed criteria set. In assessing the sustainability of any enterprise or technology, consideration should be given to at least the following:

- The sustainability (or continuity) of supply, and quality of inputs.
- The social, environmental and economic costs of providing the inputs (e.g., depletion of resources elsewhere).
- The long-term continuity (or sustainability) of production.
- Financial viability.
- Social impact and equity.
- Environmental impact.
- Efficiency of conversion of resources into useful product.

When defining sustainability, we must be clear what the objective of the definition is, and come to a conclusion of what it means to a particular concept, in this case different types of aquaculture practices. Sustainability can only be attained when environmental conditions are appropriate and maintained, and this includes ecological, socio-anthropological and economic aspects of environment. Therefore, any activity, use or practice should comprehend and include the next six general steps for sustainable resource management.

Environmental Resource Assessment

Inventories of marine, coastal and islands natural and human resources are a necessary first step for successful management programs. To enhance resource development capabilities, a country/ local community should acquire and maintain a comprehensive inventory of the physical and biological resources of the coastal area as well as their uses and users. The inventory will provide a database for making decisions about long-term goals, such as ecosystem preservation, that might conflict with the immediate development of aquaculture. This first step is necessary for assessing the coastal zone vulnerability to various activity impacts, and it provides one of the basic requirements for development of an integrated coastal resource management program. Environmental resource assessment has to include a long term and in depth research studies of organisms that are being cultured or are intended for culture, as well as understanding of utilized ecosystems bio complexity and healthy functioning.

Environmental Impact Assessment

Information about the impact that aquaculture will have on the environment must be provided in a clear form to decision-makers and stakeholders. Impact assessments should be incorporated in each phase of aquaculture development projects. It also has to present clear options for the mitigation of impacts and for environmentally sound management. Impact assessment should be based on the best available knowledge and provide timely technical information to environmental decision-makers while acknowledging uncertainties. The most important development in the environmental decision process in the last decade has been the inclusion of environmental impact assessments by regulatory managers.

Policy Framework and Regulatory Measures

This is a basic tool for training and educating, as well as for local community participation in decision-making processes. The policy statement should declare the intention of a state/nation to review and regulate, in this case aquaculture activities affecting the sustainable use of the coastal renewable natural resources. Formulation of a policy framework for coastal and marine management must address cross-sectoral issues that infringe on coastal resource management and national development planning. The basic approach is to review and analyze existing institutional and legal mechanisms (including regulations and enforcement) for integrated coastal and marine management and aquaculture development potential. Based on this review, the country should propose a generic institutional and legislative framework to address coastal issues and encourage integrating aquaculture in the coastal zone management plan. Through a series of application scenarios, the policy framework will become a basic tool for training and educating decision makers, resource managers, scientists, stakeholders, and the public in general. In addition, comprehensive policies and institutional legal frameworks should recognize the potential benefit of traditional tenure and management systems, and ensure that they are incorporated into the rules and regulations for conservation and sustainable use, where relevant.

Socio-cultural and Economic Assessment

Socio-cultural and economic evaluations are important components of a systematic assessment of coastal resources and potential aquaculture development. It provides an economic framework from which differing adaptation strategies (solutions) can be studied. In an assessment of alternatives, decision-makers should be provided with information on how each option compares in respect to the relative costs and benefits for each aquaculture impact. Identification of the full range of reasonable alternatives to resolve a conflict among competing interests means that no feasible options for maximizing benefits and minimizing costs have been missed. Therefore, successful selection of suitable locations for aquaculture and the long-term maintenance of site suitability require accurate assessment of both existing conditions and probable trends in environmental, economic, and social factors.

Implementation of Sustainable Aquaculture

The success of a coastal resource management program is based on the country's ability to understand how an effectively established program manages natural and human coastal resources. It is necessary to establish monitoring and evaluation of land use decisions and changes in coastal resources as well as in their integral uses. The basic question of an implementation strategy is how to apply science, and develop and implement best management practice (BMP) for aquaculture? Comprehensive BMP should become a 'living document', open to revisions and expansion. Established BMP provides consistent national standards and practices for implementation of different types of aquaculture in the coastal areas. They provide constant monitoring and control, strengthening environmental protection and sustainable aquaculture development in the coastal areas. Aquaculture and other coastal industries, agencies, and environmental organizations have recognized the need for BMPs. BMPs for site planning and for ICZM practices provide opportunities for early intervention and collaborative review of new activities. By publishing public standards and goals in advance of the submission of plans by a private developer, for example, one can provide guidance before major investments are made in site development. These standards (local, national, and international) also provide objective measures (indicators and criteria) that can be used by communities and environmental NGOs to question specific elements of aquaculture development proposals, as well as to award if it is environmentally sound and sustainable.

Monitoring and Evaluation of Sustainable Aquaculture

Monitoring means acquisition, management, synthesis, interpretation, and analysis of data with an emphasis on temporal and spatial scales. It should be coupled with research programs designed to improve the appropriateness of routine measurements and allow interpretations of the implications of monitoring results. A useful monitoring program provides mechanisms to ensure that knowledge is used to convert data collected into useful information. In addition, the purpose for monitoring implementation of aquaculture practices is to assure that the major policies (goals, comprehensive plans, and agency authorities) are properly implemented. Monitoring will assess the cumulative effects of changes and assure that management program elements for aquaculture are updated to reflect changing needs and circumstances, consistent with its basic requirements. This approach will provide a basis for a general evaluation of the aquaculture success or failure in achieving its overall objectives of balanced development and resource protection/ conservation.

Practitioners have discovered that sustainable aquaculture must not only maximize benefits, but also minimize accumulation of detriments, as well as other types of negative impacts on natural and social environment. One of the early lessons learned has been that increasing the density of organisms in a culture operation results in significant waste disposal problems. For example the intensive culture of carnivorous fish and prawns risks organic pollution from uneaten food or feces. This results in a potential degradation of the environment and a loss of suitability for the culture practice. Determination of suitability for aquaculture involves an evaluation of natural and anthropogenic limitations of a certain area in order to decide if the locality can support the activity. The carrying capacity for aquaculture is defined as the maximum number of users (marine species) that can be supported by a natural or man-made resource without producing negative environmental consequences to their future activity, productivity and quality.

The usual sequence in aquaculture practices has been 'development and research' rather than 'research and development', resulting in environmental damage in many cases. As more research is conducted, more information is gathered, and more case studies are evaluated, the levels of uncertainty will be reduced and our predictions should get better. Yet, we must realize that when we address a new issue (e.g., introduction of transgenics, GMOs) our knowledge base is almost nil and a huge amount of uncertainty therefore exists. Even if we do laboratory experiments, develop the best possible models and make predictions, only the introduction of transgenics and a complete study of their effects will allow us to reach some conclusions about actual effects. However, fundamental to the need for sustainability is the development of government-approved and industry/stakeholders-led, for example, 'Environmental Codes of Practice' that incorporate science-based standards and principles of adaptive risk management and that are supported by transparency and accountability in monitoring, evaluation, reporting and enforcement.

Recommendations of Sustainable Aquaculture

Based on existing national and international efforts to promote sustainable aquaculture, here is a list of key regulations and policies that should continue to be addressed, developed and implemented:

- Forming integrated coastal and rural/community management plans.
- Emphasize use of processed feeds and not fresh feeds such as low valued marine fish.
- Setting suitability standards and indicators for aquaculture (e.g., benthic fauna as bio-indicators).
- Setting monitoring and evaluation mechanisms (M&E).
- Establishing quality standards (certifications) for environmentally friendly practice, processing and sale in the aquaculture industry.
- Expanding finfish aquaculture industry should farm low trophic level fish in extensive, more responsible practices (e.g., polyculture of herbivorous carps).
- Developing integrated farming systems – integrated aquaculture–agriculture.
- Developing extensive polyculture practices.
- Eliminating government subsidies for ecologically unsound practices, and establish enforced regulatory measure for protection of coastal and ocean ecosystems (the same should apply to agriculture and fisheries).

There is a need to establish and implement environmental site suitability indicators for sustainable aquaculture development. Indicators are key variables that signal change, and can be physical, biological, chemical, social, or/and economic. They may be directly measurable or calculated from measurements of a number of data sets, or derived from other information (derived indicators). In addition, indicators can and should guide policy and help direct scientific research.

International, regional and national regulations related to environmental quality standards and indicators are well developed for the control of water quality and chemicals, although further development is required for sediment and ecosystem quality. Environmental monitoring surveys should use the best available practices and technologies for the environmental monitoring of impacts and modeling of carrying capacity at farm sites. Multiple criteria analysis must be considered when determining the suitability of a site for different types of aquaculture. Regions characterized by poor circulation, extensive accumulations of organic sediment, overwhelming recreational and commercial endeavors (marinas, race-courses, diving, fishing, port activities, etc.) will, generally, not effective culture sites. In general, the larger the number of indicators evaluated, the more comprehensive the assessment of aquaculture development potential will be. The assessment process has to include:

- Environmental assessment and EIA.
- Social and cultural heritage assessment.
- Economic development assessment.
- Land use/land cover.
- Acquire satellite images to identify each crop type, which relates to better information on pesticide pollution and negative impacts of land based activities; especially in the areas within 1 km of water.
- Use conflict assessment and analysis, taking into account current and future regional and local development plans.

2

Aquaculture: Methods and Practices

Aquaculture comprises numerous methods and practices based on what, where and how it is practiced. Mariculture, algaculture, pond culture, running water culture, recirculating aquaculture systems, rise-fish culture, raceway aquaculture, monoculture, etc. are just a few types of aquaculture. The topics elaborated in this chapter will help in gaining a better perspective about the different methods and practices associated with aquaculture.

Mariculture

Mariculture is a specialized branch of aquaculture involving the cultivation of marine organisms for food and other products in an enclosed section of the sea (cages/pens), or in tanks, ponds or raceways which are filled with saline water. Generally the mariculture production data focuses on commercial data, which is the output of marine and brackish farming activities for profit whose final harvest is used for human consumption. It is a promising sector by which the additional marine fish requirement can be met in the future years. It is also the fastest growing sub-sector of aquaculture. At global level, mariculture produces many high value finfish, crustaceans, and mollusc's viz. oysters, mussels, clams, cockles and scallops. In 2013 mariculture has contributed around 25.5 million tonnes of food fish globally which formed about 36.3 % of the food fish aquaculture production. (World food fish aquaculture production was 70.2 million tonnes in 2013). Molluscs dominated the global mariculture production (59.7%) followed by finfish (22.7%), crustaceans (16.2%) and others (1.4%). In addition about 26.9 million tonnes of macro algae and seaweeds were also produced by mariculture. The total mariculture production including the sea weeds was 52.4 million tonnes in 2013 which constituted 53.9% of the total aquaculture production during the year (The total global aquaculture production including the aquatic plants was 97.2 million tonnes in 2013).

Marine Finfish

In recent years at a global level a rapid growth in marine finfish culture is noted which has shown an average annual growth rate of 9.3% from 1990 onwards. The major finfish groups which are maricultured include salmonids, amberjacks, sea breams, sea basses, croakers, groupers, drums, mullets, turbot, other flatfishes, snappers, cobia, pompano, cods, puffers and tunas. The expansion of sea cage farming on a global basis can be attributed as a shot in the arm for the increased farming of marine finfish. Cage culture has made possible the large-scale production of commercial finfish in many parts of the world and can be considered as the most efficient and economical way of rising fish. The most vital prerequisite for the development of sea cage farming is the technology for breeding and seed production and the reliable supply of good quality hatchery produced seeds of suitable high value marine finfishes. In India, much research attention was not given for developing seed production methods for high value finfishes suited for sea farming. Till recently we had commercial level seed production technology of only one marine finfish – the Asian sea bass (Lates calcarifer). Here also private entrepreneurship for seed production has not yet been developed. Unless an intensified research on the development of commercial level seed production technologies is taken up, sea farming cannot emerge as a significant seafood production sector in the country. In the recent past, the Central Marine Fisheries Research Institute (CMFRI) has been intensifying its research activities on the breeding and seed production of high value marine finfish and success was achieved in the breeding and seed production of cobia (Rachycentron canadum) and silver pompano (Trachinotus blochii) for the first time in the country at Mandapam Regional Centre of CMFRI.

Asian Seabass

Comprehensive technology for controlled breeding of seabass was developed in 1997 and since then the technology has been further refined and validated. The technology includes captive broodstock development, acceleration of maturation, providing optimum conditions like water quality management, health management and feed management, induction of spawning through hormonal administration and facilitating natural spawning in the Recirculating Aquaculture System (RAS). Larvae are reared feeding with live feeds like rotifers up to 9th day followed by Artemia nauplii up to 20 days and afterwards weaned to formulated diet or shrimp/fish meat. The fry are further reared in nurseries. Several pond farming and cage farming demonstrations were successfully carried out.

Cobia

Fast growth rate, adaptability for captive breeding, low cost of production, good meat quality and high market demand are some of the attributes that makes cobia an excellent species for aquaculture. In recent years the seed production and farming of cobia is rapidly gaining momentum in many Asian countries. Envisaging the prospects of cobia farming in India, CMFRI has developed for the first time in the country the broodstock development, breeding and seed production of cobia and several successful seed production trials were conducted and the technology is now standardized at its Mandapam Regional Centre.

The farming protocols for the hatchery produced cobia fingerlings in sea cages with different feeding strategies were developed, tested and validated. Based on the trials, an economically viable farming model has been evolved. Several front line demonstrations and participatory farming were successfully carried out.

Silver Pompano

Among the many high value marine tropical finfish that could be farmed in India, the silver pompano, Trachinotus blochii is one of the topmost, mainly due to its fast growth rate, good meat quality and high market demand. The silver pompano is caught only sporadically in the commercial fishery and hence its availability is rather scarce. It is a much sought after species and hence the demand can only be met through aquaculture. The farming can be successfully carried out in ponds, tanks and floating sea cages. The species is pelagic, very active and is able to acclimatize and grow well even at a lower salinity of about 10 ppt and hence is suitable for farming in the vast low saline waters of our country besides its potential for sea cage farming.

CMFRI has successfully developed and standardized the broodstock development, induction of spawning, larviculture and fingerling production of silver pompano for the first time in India. The first farming demonstration from the hatchery produced seed was carried out in a coastal aquaculture pond at Antharvedi Village, East Godavari District, and Andhra Pradesh. The growth performance, survival and productive capacity of silver pompano, Trachinotus blochii, were evaluated in a brackish water pond. Based on the experience gained from the above demonstration, farming protocols were evolved. Several pond and cage farming front line demonstrations and participatory farming were successfully carried out.

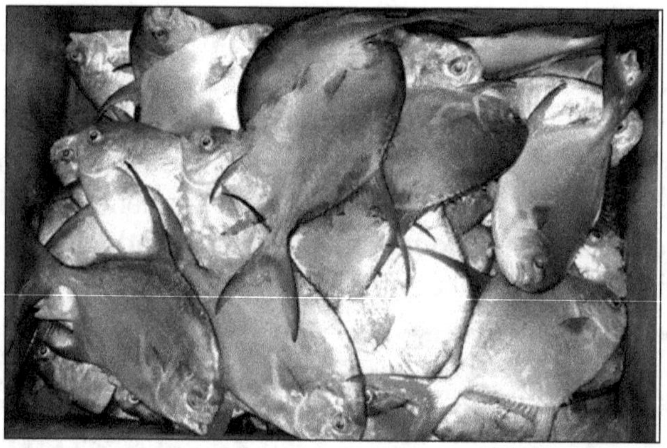

Ornamental Fish

Culture On a global basis a lucrative marine ornamental fish trade has emerged in recent years which area low volume high value industry. There are a wide variety of ornamental fishes in the coral reef ecosystems along the Indian coast, which if judiciously used, can earn a sizeable foreign exchange. However, a long term sustainable trade of marine ornamental fishes could be developed only through hatchery produced fish. The Central Marine Fisheries Research Institute has pioneered in the development of techniques for breeding, seed production and culture of more than a dozen species of marine ornamental fishes which are in heavy demand in the national and international trade. They include Amphiprion percula, A.ocellaris, A. perideraion, A. ephippium, Dascyllus aruanus, Pomacentrus caeruleus and Chrysiptera cyanea. Hatchery production and culture of marine tropical ornamental fish is very lucrative due to the high price per unit of ornamental fish. The clown fishes and damselfishes of the family Pomacentridae offer immediate scope for hatchery production due to the availability of seed production methodologies.

Sea Cage Farming

The sea cage farming has been expanding in recent years on a global basis and it is viewed by many stakeholders in the industry as the aquaculture system of the millennium. Cage culture has made possible the large-scale production of commercial finfish in many parts of the world and can be considered as the most efficient and economical way of rising fish. The rapid growth of the industry in most countries can be attributed to:

- Availability of suitable sites for cage culture.
- Well established breeding techniques that yield a sufficient quantity of various marine and freshwater fish juveniles.
- Availability of supporting industries such and feed, net manufactures, fish processors etc.
- Strong research and development initiatives from institutions, governments and universities.
- The private sector ensuring refinement and improvement of techniques/ culture systems, thereby further developing the industry.

Total reported cage aquaculture production from 62 countries and provinces/regions from where data is available amounted to 2412167 tonnes (excluding China) On the basis of the reported information, the major cage culture producers in 2005 included - Norway (652306 tonnes), Chile (588 060 tonnes), Japan (272 821 tonnes), United Kingdom (135 253 tonnes), Vietnam (126 000 tonnes), Greece (76 577 tonnes), Turkey (78 924 tonnes), and the Philippines (66 249 tonnes). Currently on a global basis commercial cage culture has been restricted to the culture of high value, compound feed fed finfish species, including salmon, Japanese amberjack, red sea bream, yellow croaker, European sea bass, gilthead sea bream, cobia and groupers. Cage culture systems employed by farmers vary from traditional family owned cage farms (Asian countries) to modern commercial large scale salmon and trout cage farms in Northern Europe and Americas.

Marine cage farming is relatively new in Asia and was developed initially in Japan for species such as yellowtail (Seriola quinqueradiata) and red sea bream Pagrus major. Over the last twenty years the cage farming practice has spread almost throughout Asia. The major cage farming countries are China, Indonesia, Taiwan Province of China and Vietnam. A large number of finfish species are farmed in cages in Asia viz. groupers, snappers, carangids, seabass and cobia. In most countries individual operations are not large, and often a clustering of farming activities, which is due to limited site availability in coastal waters, is seen.

When compared to many countries in the Asia-Pacific Region, India is still in its infancy in sea cage farming. For the first time in India as part of R & D a marine cage of 15 m diameter with HDPE frame was successfully launched in 2007 and operated at Visakhapatnam, in the east coast of India by the Central Marine Fisheries Research Institute. Since then, a lot of innovations on designing and fabrication of cages and mooring systems were made which led to the development of better designs of cages of 6m diameter with improved mooring systems that can withstand rough sea conditions. Subsequently demonstrations of cage farming were undertaken along different parts of the Indian coast under a participatory mode with the local coastal fishermen. Successful sea cage farming demonstrations were conducted at Kanyakumari, Vizhinjam, Kochi, Mangalore, Karwar, Veraval, Mandapam, Chennai and Balasore. Cobia, Sea bass and spiny lobsters were the major groups employed for farming. These demonstrations have created an awareness regarding the prospects of sea cage farming in India. Many entrepreneurs, fishermen and farmers are coming forward to take up this venture.

The commercial cage farming of cobia undertaken by Cobia Fisherman Welfare Association, a self-help group from Rameswaram under the technical support of Mandapam Regional Centre of CMFRI is a step forward in the popularization of sea cage farming in India. Ten cages of 6m diameter and 3.5m depth were fabricated and floated by them. All the investments in the fabrication of the cages, the cost of seeds, feeds and managing the sea cage farm were borne by the association. A total of 6400 fingerlings of hatchery produced cobia were supplied from Mandapam Regional Centre. The farming was initiated during the middle of November 2013. The harvest of cobia was conducted on 8th May 2014. The length of fish harvest ranged from 48 to 62 cm and weight from 1.0 to 2.3 kg. The farm gate price was 270 / kg. A total of 10 tonnes of fish was harvested during the fishing ban period and yielded a good farm gate price. This has created wide spread interest among fishermen communities for taking up sea cage farming in the area. Several successful participatory cage farming demonstrations taken up at different parts of Indian coast is paving the way for the spreading of cage farming and in the immediate future it can contribute to additional sea food production.

Mussel Farming

In the wild, mussels are mostly found in the littoral and sublittoral zone in clusters on various substrates. The two mussel species found along Indian seas are Perna virdis and Perna indica. Perna viridis is distributed along the north and south of east coast and the south west coast of India. The brown mussel Perna indica species is found along the south west coast. The Institute has developed technologies for culture of bivalve's viz. raft method (in bays, inshore waters), rack method (in brackish water, estuaries) or long line method (open sea). These methods are commonly adopted for mussel farming (Perna indica and P. viridis). Mussel seeds of 15-25 mm size collected from intertidal and sub tidal beds are attached to coir/nylon ropes of 1-6 m length and enveloped by mosquito or cotton netting. Seeds get attached to rope within a few days while the netting disintegrates. The seeded ropes are hung from rafts, racks or long lines. A harvestable size of 70-80 mm is reached in 5-7 months and production of 12-14 kg mussel (shell on) per metre of rope can be obtained. Innovations such as automatic seeding machines and depuration protocols were also evolved. The farming of mussels is currently being practiced commercially at brackish water areas of Malabar Coast, Kerala.

When mussel farming is taken up on a larger scale, depending on the wild seed may not be practicable because it may affect the wild mussel fisheries. Hence low value high volume seed production technologies are needed for hatchery production of mussel seeds for farming. CMFRI has already succeeded in the low cost production of mussel seeds in hatchery and standardization of the protocols for large-scale production is needed.

Edible Oyster Farming

CMFRI has developed methods for edible oyster (Crassostrea madrasensis) culture and has produced a complete package of technology, which is presently being widely adopted by small scale farmers in shallow estuaries, bays and backwaters. In the adopted rack and ren method, a series of vertical poles are driven into the bottom in rows, on top of which horizontal bars are placed. Spat collection is done mainly from the wild on suitable cultch materials. Spat collectors consist of clean oyster shells (5-6 Nos.) suspended on a 3 mm nylon rope at spaced intervals of 15-20 cm and suspended from racks, close to natural oyster beds. Spat collection and further rearing is carried out at the same farm site and harvestable size of 80 mm is reached in 8-10 months. Harvesting is done manually with a production rate of 8-10 tonnes/ha. Oyster shells are also in demand by local cement and lime industry. Hatchery production technique for edible oyster seeds is also developed by CMFRI.

Molluscan shellfish (mussels, oysters and clams) are much sought after and widely consumed throughout the world as gourmet food. But in India this nutritious seafood have not found much acceptance. Currently farmed mussel and oyster production in India is between 10,000 and 15,000 tonnes. In order to create awareness on the general public on molluscan shellfish as a highly nutritious food, ShellCon 2014 was organised by CMFRI in which shellfish food festival and programmes for the popularization of the consumption of molluscan shellfish were conducted.

Mabe Pearl Production

CMFRI successfully developed and standardized a simple technique for value added marine pearls, called mabe pearls. A mabe pearl is a dome shaped or image pearl produced by placing a miniature image against the side of the oyster shell interior. The result is an exquisite pearly nacre coated image. The main advantage is the very short gestation period (2 months) and the superior quality of the nacre of Indian pearl oyster Pinctada fucata. There are two ways to make pearl oyster spat available for pearl oyster farming;

- Hatchery production of seed.
- Wild collection of seed by setting up artificial spat collectors in the sea at subsurface during oyster spawning season.

The hatchery technology developed by CMFRI is helpful in overcoming the problem of insufficient supply of oysters for cultured pearl production.

Seaweed Culture

Around 60 species of commercially important seaweeds with a standing crop of one lakh tonne occur along the Indian coast. Seaweed products like agar, algin, carrageenan and liquid fertilizer are in demand in global markets and some economically viable seaweed cultivation technologies have been developed in India. CMFRI has developed technology to culture seaweeds by either vegetative propagation using fragments of seaweeds collected from natural beds or spores (tetraspores/carpospores). It has the potential to develop in our large productive coastal belts. The rate of production of Gelidiella acerosa from culture amounts to 5 tonnes dry weight per hectare, while

Gracilaria edulis and Hypnea production is about 15 tonnes dry weight per hectare. Recently the culture of the carageenan yielding sea weed Kappaphycus alvarezii has become very popular due to its fast growth and less susceptibility to grazing by fishes and is being cultivated extensively along the Ramanathapuram, Pudukkottai, Tanjore, Tuticorin and Kanyakumari districts of Tamil Nadu producing about 17000 t wet weight annually.

Integrated Multitrophic Aquaculture (IMTA)

On a global basis, the mariculture practices are dominated by intensive monocultures which have led to sustainability problems, environmental degradation and consequent disease problems. In this context, the idea of bio-mitigation of the environment along with increased biomass production integrating commercially important species of different trophic levels is emerging as an innovation in aquaculture. Integrated Multi trophic aquaculture (IMTA) is the practice which combines in appropriate proportions the cultivation of fed aquaculture species (E.g. fin fish / shrimp) with organic extractive aquaculture species (e.g. shell / herbivorous fish) and inorganic extractive aquaculture species (e.g. seaweed) to create balanced systems for environmental stability (bio mitigation) economic stability (product diversification and risk reduction) and social acceptability (better management practices). IMTA is well recognized as a mitigation approach against the excess nutrients / organic matter generated by intensive aquaculture activities especially in marine waters, since it incorporates species from different trophic levels in the same system. In addition, it is also relevant in the implementation of ecosystem approach to aquaculture (EAA) propagated by FAO. IMTA can also increase the production capacity of a particular site. It is well understood that the increasing use of coastal waters worldwide coupled with rapid growth and expansion of mariculture demand for more sustainable practices and hence the concept of IMTA has much relevance and scope.

Seed availability is the major constraint for the initiation of commercial level farming of marine finfishes and shellfishes. The huge demand for cobia and pompano seeds received at CMFRI from fish farmers and entrepreneurs is indicative of the need of the sector. Hence there is an urgent need to establish marine finfish hatcheries by fisheries development agencies /private sector to ensure the seed availability. In addition, it is required to intensify research programmes for the development of seed production techniques for at least one dozen species of high value marine fishes. In this context, CMFRI has already taken up broodstock development and seed production of orange spotted Grouper Epinephelus coioides, Indian Pompano Trachinotus mookalee and Malabar Red Snapper Lutjanus argentimaculatus. Initial success has already been obtained in the broodstock development and seed production of E. coioides and T. mookalee at the Vishakapatnam Research Centre of CMFRI. Broodstock development of L. argentimaculatus is being pursued. If seed production technologies of more species are available, the farmers will be able to select the species as per the need of the locality.

The commercial level farming of lucrative shellfish species like the sand lobster, Thenus unimaculatus and the blue swimmer crab Portunus pelagicus can also be practiced if hatchery produced seeds are available. CMFRI is able to succeed in the seed production of both the species and now research is being focused on the standardization of these techniques for consistent production. Similarly seed production techniques are already developed by the Institute for edible oyster, pearl oyster and green mussel. The methods can be scaled to commercial level production as per the requirement of the sector.

The development of farming systems especially the sea cage farming deserves prime attention. To promote sea cage farming in the country, identification of suitable sites with proper depth, water quality and water current are required. Site selection survey and identification of at least a dozen sites suitable for cage farming by the entrepreneurs and farmers deserves urgent attention. Availability of logistic support for cage farming should be given careful consideration if a profitable business is to be established. Cage farming has to be promoted away from the human settlements, discharge points of industrial and municipal waste, so as to maintain ideal water quality for sea farming. Further, policy for leasing the suitable sites, bank finance, and governmental support through subsidy assistance are the needs of the hour.

The bivalve farming which is already being practiced at a few locations can be further expanded. The carrying capacity assessment, low value availability of hatchery produced seed and the feasibility of open sea farming of bivalves require attention by the R&D sector. Similarly the expansion of sea weed farming offers immense scope. A concerted effort by the developmental agencies for popularization of sea weed farming is warranted.

On a global basis, the mariculture practices are dominated by intensive monocultures which have led to sustainability problems, environmental degradation and consequent disease problems. In this regard, the bio mitigation impact of IMTA is away forward in sustainable aquaculture. In addition, it is also relevant in the implementation of ecosystem approach to aquaculture (EAA) propagated by FAO. The development of IMTA in marine and coastal environments has not been demonstrated as a viable enterprise in India and hence there is an urgent need to impart front line demonstrations on this potential sector of mariculture to different stake holders.

The development of commercial level seed production technologies for a few species of high market value finfish and shellfish, establishment of hatcheries by fisheries development agencies, identification of appropriate cage/coastal farming sites, development of economically viable farming protocols, formulation of suitable grow-out feeds, health management protocols, development of mariculture policies and appropriate marketing strategies can go a long way to promote mariculture as a substantial contributor of sea food production.

Effects of Mariculture on Marine Environment

Aquaculture activities are increasing worldwide due to the rapid increase in world population and increase in demand for aquaculture products. Increased demand on fish is mainly satisfied by salmon (Salmo salar) in cold regions, by sea bass (Dicentrarchus labrax) and sea bream.

Aquaculture industry is growing continuously here at the environmental effects of the aquaculture activities are reached remarkable size. Aquatic environment is getting more polluted biologically, physically and chemically also changing appearance of natural environment. It is hard to distinguish the real and potential effects of the aquaculture or determine the enormity. To deal with the rapid growth of the mariculture industry, means that public administrators deciding not to regulate growth of such an industry must take risks. Seas are getting polluted by direct waste disposal and by chemical use. Impacts of the discarded waste in water column depend on geographical features of the regions and bottom features of the sea. Dissolved constituents (ammonium, phosphorus and nitrogen dissolved organic carbons), oils and lipids released from the diet can be listed as some of the discarded wastes.

Impacts of Mari Culture on Water Quality

Effects of waste on water body depend on the key features of the habitat, waste quantity, time-scale over which the releases take place and characteristics of the water column. Aquaculture originated wastes can be subdivided into two major groups; solid waste and soluble waste. Solid wastes are composed of uneaten pellet feed, faeces, fish scale and mucus. Since most of them are denser than water, they sink to the bottom. Soluble wastes are ammonium, urea and some of the particles that solved from solid wastes. Among these wastes, nitrogen, phosphate groups and suspended substances are the main components that affect water quality. Moreover, some of the vitamins and minerals (trace elements) are effective on algae populations. Aquaculture originated pollution is generally divided into two categories as organic and chemical pollution.

Organic Pollution

The main sources of organic pollutants are fish faeces, uneaten feed and dead individuals. Faeces produced by fish are related with the feed ingredients. 25-50% of the consumed feeds are excreted to environment as faeces. Intensive fish farming causes large amounts of organic waste accumulation in the sediments and water column. A sediment collector placed under salmon cultivation cages collects 14.7-52 kg of organic waste in each square meter per annum, which reveals the extent of organic pollution originating from aquaculture activities. Aquaculture activities lead a number of ecological issues. For instance, aquaculture facilities such as cage systems significantly reduce water current speed thus cause accumulation of organic substances in surrounding environment. Decomposition of these organic wastes leads deterioration of water quality and increase water oxygen requirement. Negative effects like turbidity, accumulation of organic waste in the sediment, formation of anoxic zone near sea floor, accumulation of toxic substances and spread of diseases are influential in areas where aquaculture is in practice, especially in restricted exchange environments.

Formation of gasses such as H_2S (hydrogen sulfide), CH_4 (Methane) and CO_2 (Carbon dioxide) on sediment slows down growth rate of living organisms in the local area and makes them more vulnerable to diseases. Water oxygen level is reduced by direct decomposition of organic waste originated from aquaculture activities. Water quality changes by releasing nutrients like nitrogen and phosphate into the water column. Excessive amount of organic nutrients promote growth of water plants and algae. When toxin producing algae grow excessively, they may reach blooming concentrations and cause negative impacts to other organisms. Cultured fish may die off due to the disorders occurring in the gills and oxygen deficiency as a result of water quality changes related with sudden growth of algae. Consumers of bio toxin contaminated shellfish can get poisoned and die as well. When algae die in large numbers, dead organic substances start to decay and this process consumes oxygen, leads to hypoxia. Accumulation of organic and inorganic wastes in areas with low water exchange rate may lead hyper nitrification as well.

In this sense, permanent aquaculture activities in such regions accelerate accumulation of waste. On the other hand, aquaculture activities held in deep waters or areas with high bottom current rate may result in wide spread of wastes. Let's make a comparison to explain extend of pollution caused by aquaculture activities. An average person excretes 4kg of nitrogen and 1.1kg of phosphate per year. For every ton of salmon 55kg of nitrogen and 4.8kg of phosphate are released into the water column through faeces per year. A salmon cultivation facility with a production capacity of50000 tons a year release an amount of nitrogen equals to that of 682000 people, and an amount of phosphate equals to that of 216000 people into the water in a year.

Eutrophication

Eutrophication simply means enhanced nourishment and refers to the stimulation of phytoplankton or macro fit (Algae and plants) growth in aquatic environment. To date, in all modeling studies, only effects of mariculture originated nutrient waste on phytoplankton growth were investigated. Especially, phytoplankton growth in summer months controlled by nutritious elements (Nitrogen, phosphate etc.) was modeled. Factors of this kind of model may lineup as; determination of bottom structure, water column affected from waste and dilution rate, calculation of nutrient level and determination the relationship between phytoplankton growth and nutrients.

In Baltic Sea, relations between sea water sensitivity, (depends on the interaction between open water and coastal water) nutrient concentration, phytoplankton biomass and secchi depth compared by using multiple regressions techniques. Depending on the sensitivity of the coastal areas, while phytoplankton biomass increases, secchi depth decreases. But since the water in shallow areas is under influence of wave induced re suspension, secchi depth method is unreliable. Therefore, the most accurate method for determination of the phytoplankton biomass is the measurement of chlorophyll content. In another study, direct and indirect effect of eutrophication was investigated between 2001-2011 in Baltic Sea by using nutrient concentration, chlorophyll-a, and secchi depth indicators. Study revealed that entire open Baltic Sea was affected by eutrophication between 2007 and 2011.

Chemical Pollution

A variety of chemicals used for disinfection and disease outbreak treatment in aquaculture. Environmental impacts of chemical compounds depend on toxic effect of the compound, pathogens sensitivity to antibiotics, therapeutics and active time zone of the chemical in the environment. Most of the chemicals were adapted from the poultry and cattle husbandry and their possible effect on the marine environment isn't fully investigated.

Chemicals in use in mariculture can be classified as disinfectants, anti foulings and veterinary medicines. Main veterinary medicines are antibiotics, anesthetics and pesticides. Veterinary medicines are widely used in fish farms but it is not so common in crustacean and bivalve cultures. Most of the chemical treatment scenarios in Mariculture result in direct release of chemicals into water column and sediment. For the cage systems Atlantic salmon, rainbow trout, sea bass and sea bream are the most common fish species and most of the chemicals used for which cause pollution. Among numerous definitions of marine pollution the most common one is GESAMP's (Joint Group of Experts on the Scientific Aspects of Marine Pollution), which states that marine pollution is "the damage to living creatures in the environment, the hindrance to marine activities including fishery, the inability to use sea water and the reduction of amenities as a result of changes in environmental conditions due to deleterious effects induced by various waste materials and/or energy left to the marine environment (including estuaries) by people directly or indirectly.

When a disease emerges, farmers try to control the outbreak by using antibiotics. As a result of that, antibiotic makes its way through water by uneaten antibiotic treated feed or feaces. In order to control external parasites such as sea lice, different kinds of pesticides are used. Pesticides could be given to fish via feed or by bath. Pesticides such as ivermectin have high toxic effects on marine organisms. Bio fouling is one of the major problems in cage farming. Proliferation of bio fouling organisms adds considerable weight to open water cage net which threats the stability of the structure and readily reduce water flow through which reduces water quality within the cage systems. Anti-fouling paints and coatings are used in order to slow growth of such organisms on net to maintain water flow through to systems. Active matter of such paints is generally copper. Degradation of the paint in time causes copper to penetrate water, which poses toxic effects on marine creatures. Another toxic metal originated from aquaculture is zinc. Zinc sulfate ($ZnSO_4$) is used as feed additive to prevent disease outbreaks.

In 2001, government of Canada surveyed surrounding waters of the salmon cultures and find out that copper and zinc concentrations are higher than legal limit in 48% of the cage systems. 0.15 mg/L copper is lethal for shrimps. For urchin 1.4-11.4 ppm copper and 0.327 ppm zinc is lethal. For amphipoda, 0.8ppm zinc causes 23% mortality. When both copper and zinc present together in the environment their synergistic toxic effect becomes greater than the sum of their individual effects. Metals generally have sub lethal effect on aquatic organisms. Copper and zinc toxicity inhibit chemo receptors and damage digestion system. Acute exposure to copper inhibits both uptake and excretion of ammonia in rainbow trout. It also makes the organism an easy prey for predators by disrupting their nervous system.

Antibiotics

Chemicals have wide range of usage in fish health management. Report prepared by GESAP and WHO (World Health Organization) mention chemical usage pose a threat to human health. Despite the reduction in antibiotic and organophosphate use in aquaculture, synthetic parathyroid, colorant, preservatives, anti-parasitic and other chemicals that cause marine pollution are still in use. Use of chemicals not only affects marine environment but also pose a risk to the farmers as well. Indiscriminate use of antibiotics for fish health management is one of the major problems in aquaculture. Un-metobolized antibiotics are often passed to the aquatic environment by uneaten antibiotic treated feed and by faeces. In both water column and sediment, different classes of antibiotics have different half-life. Oxytetracycline which is one of the most common antibiotics can be effective up to 30 days in both water and sediment. In fact, oxytetracycline is still measurable on sediment after several months without being biologically active. In cage systems, chemicals in uneaten antibiotic treated feed and faeces scatter to the surrounding environment by waves and current. Wild fish and filter feeders such as mussels and oysters consume antibiotic treated feed or it passes to sediment.

Bioactive substances such as antibiotics and pesticides are generally used for controlling disease and parasite in aquaculture. Success or failure in aquaculture depends on the correct usage of chemicals, usage duration and usage conditions against infectious diseases and parasites. Effects of bioactive matters on environment can be aligned as:

- Substances with inhibitory effect last long in animal tissue.
- Release of substances with inhibitory effect to the aquatic environment.
- Development and transmission of the antibiotic resistance gene in microorganisms.

Substances with Inhibitory Affects Last Long in Animal Tissue

Bioactive substances last long in living cells more than expected. For instance, trimethoprim in used in disease treatment, can be detectible in a mussel after 77 days. In order to totally eliminate oxytetracycline and other sulfonamide based antibiotics from trout body above 10°C, at least 60 days should pass after the last usage. Frequent use of antibiotics to prevent or treat bacterial fish diseases, may result environmental bacteria and intestinal bacteria to gain antibiotic resistance. Antibiotics make their way to muscles through digestion system. The emergence of antibiotic resistance bacteria as a result of massive use of chloramphenicol was determined in trout farms, in Italy. Antibiotic taken into body through digestion emits from intestines and passes to vascular system, from where is passes to muscles. Every antibiotic has a different withdrawal period. Since the muscle tissue of the fish carries antibiotics for a specific period of time after the treatment, fish should be kept in water for a while depending on the water temperature in order to eliminate the antibiotic residue from the tissues before putting fish on market. For instance, fish treated with oxytetracycline shouldn't be offered for sale at least for 20 days. Especially in summer months along with temperature rise, mortalities in market size fish force farmers to use antibiotics. But farmers generally don't heed withdrawal time procedures. These antibiotic containing fish offered to market before the withdrawal period pass to the consumer causing bacteria in human body gain antibiotic resistance.

Releasement of Substances with Inhibitory effect to the Aquatic Environment

Frequent use of the inhibitor substances in aquaculture and replacement of potential bioactive substances to the aquatic environment is frightening. Only 20-30 percent of the antibiotic treated feed is eaten by fish, whereas 70-80 % of it passes directly to the aquatic environment. Oxytetracycline decay fast in marine environment but most of them bind to the tiny particles and accumulate on the sediment. Accumulated oxytetracycline could still exist at enough concentration to exhibit activity even after 12 weeks. Sediment that contains antibiotics affects the marine fauna negatively. For example, oxytetracycline residue was detected in blue mussel (Mytillus edulis) 80 meters away from cages where the antibiotic was used.

While the antibiotics on the sediment surface lose its effect easily, antibiotics like oxytetracycline, oxolinic acid, sarafloxacin and flumequin within the sediment can be effective for up to 180 days. Antibiotics on the sediment become ineffective so easily not because of the degradation, but due to the spread. Antibiotics like sulfadiazine and trimethoprim have relatively less half-life within the sediment which is approximately 90 days. In the other hand, half-life of the florphenicol in sediment was calculated as 4.5 days. Various natural and synthetic chemicals, such as dichlorvos, malachite green and derris root are used in large amounts in mariculture worldwide. We might give an example in order to reveal total amount of chemical usage; in 1989, Norwegian farmers used 3488 kg of dichlorvos in order to control sea lice (Lepeophtheirus salmonis) outbreak. Disclorvos and most of the used chemicals cause serious impacts on environment and they are needed to be used wisely.

Development and Transmission of the Antibiotic Resistance Gene in Microorganisms

Antibiotics are widely used for controlling and treatment of both human and fish diseases. The amount of antibiotic use in aquaculture and in human health management is nearly same.

Thus, frequent usage of antibiotic cause development of antibiotic resistance gene in both human and animal pathogens and treatment of the bacterial diseases with antibiotics is getting harder and harder each day. Antibiotics are not only using as disease controller but also as growth and feed conversion promoters. In 2000, Denmark banned all non-therapeutic uses of antibiotics. In 2006, EU banned the use of antibiotics as growth promoter. Antibiotics used at low doses increase the development of antibiotic resistance genes in human pathogens and opportunist environmental bacteria and ease the path of gaining multiple antibiotic resistances of bacteria. Plasmids (extra chromosomal elements) located in plasma of the bacteria carry resistance genes. Development of the antibiotic resistance gene arises from the changes in plasmid DNA and chromosomal DNA mutation.

Resistance plasmids (resistance, R) can be transferred between bacteria spp. Plasmids are transferred between bacteria in two different ways; Transduction and conjugation. Most of the bacteria species including pathogenic bacteria have plasmids. Transfer of R Plasmids from resistant bacteria to sensitive bacteria causes different bacteria species develop antibiotic resistance gene. Being transferable and independently replicate able plasmids are the main factor of wide spread of the resistance genes. Since the excessive use of antibiotics in aquaculture could promote development of antibiotic resistance genes in pathogens, most of the antibiotic treatments are inefficient. Roughly 16 different antibiotics are commonly in use in aquaculture. Ampicillin, florphenicol, erythromycin neomycin, oxytetracycline and tetracycline are among them. Some of the bacteria gain resistance against these antibiotics inevitably. In 1981, resistance of Aeromonas salmonicida the etiological agent of the furunculusis disease to oxitetracycline and was 4%, resistance ratio reached 50% in 1990. Presence of antibiotic resistance gene in the environment reduces effectiveness of the antibiotic. Usage of the antibiotics as a growth promoter leads development of antibiotic resistance in human pathogen bacteria and increases the risk of antibiotic resistance gene transfer. Transfer of the R Plasmids from aquatic microorganisms to human pathogens poses a potential threat to human health.

Impact on Microbial Communities

Our understanding of the effect of mariculture on natural microbial communities is extremely Iimited, despite the fact that several marine microbes are potentially pathogenic to the cultured organisms, the indigenous biota or man. Most microbial work done in connection with mariculture has been directed towards Identification of Those organisms producing disease in cultured finfish and towards the development of Therapeutic methods to prevent their spread. In comparison, much less work has been done in the area of microbiota ecology, and specifically, Identifying how allure operations modify the environment so as to change the number and type of bacteria. The majority of bacteria in both freshwater and marine coastal and estuarine waters are heterotrophies, and the number of heterotrophs is largely determined by the concentration of organic matter. Thus, the input of organic matter associated with mariculture, and particularly the enrichment of bloom sediments, could reasonably be expected to lead to an increased number of cultivable bacteria in The vicinity of culture sites and to result into selective enhancement of specific bacterial groups.

The effect of culture activities on coliform and specifically, faecal coliform bacteria Is of particular Interest because of the importance of this group in water quality monitoring. Finfish authorities have found elevated concentrations of coliform and faecal streptococci baeterla in effluents from freshwater trout ponds. Subsequent investigations found that the fish Intestines and the sediments of the mud-bloom ponds can serve as multiplication sites for Enterobacteriaceae (E. coli, Entero and Citrobacter), Aeromonas and faecal streptococci which enter the culture systems through the Influent water and/or feed. The total number of indicator bacteria in the effluents was low, but detectable in the receiving waters lf oysters were cultivated in the effluent of fish farms they accumulated the Enterobacteriaceae, The faecal Streptococci, and also Aeromonas. Increased 1olal coliform concentrations have also been reported in receiving waters near several freshwater hatcheries in the western United States, although a large freshwater trout farm in Norway caused no change in faecal coliform numbers in receiving water. Near-surface waters near marine salmon net-cages In Sechelt Inlet, British Columbia had a higher local coliform concentration than comparable reference areas, although the increase was not statistically significant. We are In need of a clearer differentiation between faecal and local coliform, because The counts obtained with methods usually employed to evaluate sewage effluents and waste waters originating from human activities, may not adequately reflect The species and type composition encountered In fish farm effluents. The development of sound hygiene criteria, specifically adapted to the marine environment is urgently needed.

The most extensive work on microbial dynamics of fish farms is that of Austin and Allen-Austin. Freshwater trout rearing facilities have been shown 1o have Iive effect on the bacteria populations of the receiving waters in most instances bacteria numbers (CFU, - Colony Forming Units) from the effluent from these facilities are comparable or less than numbers in the influent. Alter Investigation of 6 farms. In only 8 out of 29 comparisons was the number of bacterial in the effluent substantially higher (two-fold or more) than the number in the influent. There were also live effects of culture on bacterial composition, with the taxa of the effluent generally typical of unpolluted fresh water. The most significant effect of culture was the selection for antibiotic resistant strains during chemotherapy. The use of oxolinic acid potentiated sulphonamide and oxytetracycline all increased the proportion of resistant bacteria in the effluent. This phenomenon was short-Iived, however, for after 9 days or less following cessation of treatment, the proportion of antibiotic resistant bacteria In the effluent returned to pre-treatment levels.

Austin concluded that unlike freshwater finfish culture, marine culture may result in a net contribution of certain types of bacteria into the receiving water, although it should be recognized that this conclusion is based on studies at a single farm site and may be more representative of facility management rather than a general pattern. In each of four sampling periods, the number of bacteria in the effluent from a coastal turbot rearing facility was substantially greater than the number in the Influent, with the increase ranging from 3 to 50-fold mainly indicating growth of naturally occurring bacteria. There were also compositional enhances in the bacteria including a decrease in the proportion of Acinetobacter calcoaceticus, Hyphomicrobium, Hyphomonas, Micrococcus, Photobacterium, Pseudomonas, Staphyloccus, Vibrio sp. and a commensurate increase in Bacillus spp., corynelorms Enterobacteriaceae and Prosthecomicroblum spp. in the effluent. The Gram-positive bacteria are normally associated with sediments, and the Enterobacteriaceae with faecal pollution.

It is not known whether changes in bacterial composition or numbers In the vicinity of a mariculture pose an increased risk of disease in indigenous biota. Several investigators have raised the possibility of disease transmission from cultured to wild fish, and questioned whether fish pathogens in the effluent from mariculture facilities might infect wild stocks in the vicinity. There are cases of mariculture introducing a disease organism into an area where it had not previously occurred, eventually leading to the infection of wild fish. There are, however, no documented cases of mariculture leading to increased Incidence of disease in wild fish by a pathogen already present in the environment, but demonstrating no clinical symptoms. Many fish diseases are associated with some form of stress; including vibriosis, entire red mouth, and fin rot. Deterioration in water quality, poor husbandry or other stresses imposed on fish In the culture environment can increase their susceptibility to diseases caused by opportunistic pathogens, but appearance of the disease in culture or even occurrence of the bacterial pathogens in the effluent does not necessarily pose a risk to healthy, unstressed wild fish.

Studies on bacterial growth and species composition on molluscan culture sites are few. Recently, a Swedish study found no significant differences in bacterial counts and nutrients at sampling point outside and inside a mussel farm, which was located in a well-protected Inlet with strong tidal currents (up to 10 cm/sec). Since this study was undertaken in April when water temperature is relatively low, the situation may be different during the most productive summer period.

Impacts of Mariculture on Marine Algae

Mariculture activity could bring about changes in the natural population of marine algae (phytoplankton and macroalgae) in the vicinity of the farm or at a distance where fiushing action is vigorous. These effects include stimulation of primary production, changes in the species composition of the phytoplankton reduction of phytoplankton standing crop senescence of phytoplankton blooms with subsequent localized areas of low dissolved oxygen and enhancement of macrophyle growth. These effects have Implications for public health, natural populations of marine organisms and the viability of mariculture industries.

Finfish Culture

By-products of fish metabolism and lead leachates represent a source of nutrients for phytoplankton growth and could give suitable hydrographic conditions stimulate primary production of phytoplankton of nutrients are controlling growth at the time. The Input of soluble waste into the water column will alter the natural composition of macro- and micronutrients and It has been suggested that this can bring about changes in the composition of the phytoplankton. Organic waste from fish farms might also play a role in stimulating the growth of specific algal species. For example fish farm by-products (fish flesh and faeces) enhanced the growth of at least one red-tide forming dinoflagellate. Biotin has been implicated in the effects of Gyrodinium striatum and vitamin B_{12} Is a growth requirement for the toxic micro flagellate Prymnesium parvum. Biotin and vitamin B_{12} are constituents of fish feed but their fate in the marine environment Is poorly understood. Recent British studies showed that Vitamin B_{12} in fish feed made farmed fish more susceptible to Gymnodinium attack. Gymnodinium is as such not toxic to fish but affects mainly the oxygen transfer across gills through sloughing of gill membranes. There are three mechanisms by which phytoplankton blooms might become a problem for mariculturists:

- Depletion of dissolved oxygen (DO) in and around culture facilities: During dense blooms oxygen consumption by algae during the night might reduce the concentration of DO to a level at which cultivated fish are asphyxiated. Reduction in oo during the senescence and anaerobic microbial degradation of large phytoplankon blooms might also have similar effects.

- Physical damage to sensitive tissues: Blooms of chain-forming species of Chaeloceros are known to have caused physical damage to the gills of farmed fish. It seems Likely, However, that many species of diatoms which have long spines and some of the large armored dinoflagellates could have similar effects.

- Formation of algal toxins and subsequent uptake and accumulation by cultured organisms or direct toxic effects on the cultured organisms. Shellfish toxicity (PSP and OSP) is a serious problem in many ICES member countries. Toxic algal blooms are known to have killed wild fish and have been implicated in mass mortalities of farmed fish in European waters and shellfish.

The formation of anoxic sediments beneath mariculture operations could provide suitable conditions for the accumulation of dinoflagellate cysts by preventing excystment. Dispersal of this sediment, by natural causes or for husbandry reasons could result In the release of cysts into the water column and under suitable hydrographic conditions may result in the development of a bloom. Some coastal regions might be areas of natural cyst accumulations and mariculture In these areas might be inappropriate.

The possible impact of mariculture on macro algae is not well understood. The importation of exotic species, particularly shellfish, caused the introduction of new algae which can be considered as pests. Their growth tends to replace some native algae. Development of mariculture (fish and shellfish) increases the release of nutrients (such as ammonia and phosphorus) and provides often additional substrate for attachment (e.g. cages. nets. racks. etc.). As a consequence, a large growth of green macro algae can cause increased fouling on various substrates. The widespread dispersal of their spores can induce ·green tides· near the culture installations, particularly in enclosed bays, in marshes and in ponds. These algae act in competition for nutrients with phytoplankton stocks, and this can have Implications for the cultivation of molluscs. "Green tides· can also cause large decreases in dissolved oxygen.

Shellfish Culture

Extensive culture of bivalves may affect phytoplankton levels in one of two ways. First, bivalve culture will decrease the amount of phytoplankton. Imai clearly demonstrated that the culture of 50,000 to 90,000 oysters reduced concentrating seston (dominated by phytoplankton) by between 76 to 95%. Furthermore there is laboratory evidence that the filtering efficiency of bivalves (mussels) can be so great as to effect the growth of other mussels downstream. Attempts have been made to estimate the seston requirements for growth of mussels and Japanese oysters.

Secondly, the stimulation of primary production by filter feeding bivalves has also been predicted. Campos and Marino have suggested that phytoplankton productivity would be stimulated by an increase in the rate of nutrient cycling. A number of investigations have demonstrated that recycled organic and mineral nitrogen excreted by oysters and clams is utilized by several species of phytoplankton. Arakawa was able to correlate the frequency of phytoplankton blooms In Hiroshima Bay with historical trends in oyster production in those waters. However, field evidence of increased primary production in the vicinity of shellfish culture is lacking.

Impacts of Mariculture on Sediment and Benthos

Mariculture structures can influence sedimentation simply by modification of current velocity and direction. Like any other structure placed in a moving fluid, mariculture structures may alter the fiow regime creating zones of both erosion and deposition. In areas of intensive culture, the effects of bathymetry or even coastline topography can be significant. Hanging and stick shellfish culture In the Philippines has increased sedimentation to the point that procreation of coastlines has occurred in some areas. Some of the well-studied examples are the mollusc culture grounds of France. Oyster farmers using metal racks in the intertidal zone have been confronted with sediment accumulation beneath and between culture racks. Concurrent with elevation of the sediment surface, erosion around the rack legs has accelerated sinking of the structures. Similar problems have been faced by mussel culturists using the bought technique. The numerous poles used to support the mussel reduced current velocity by up to 50%, thus accelerating the deposition of fine sediments. It is estimated that 30% of the oyster end mussel farms of France face problems of active sedimentation, forcing occasional relocation and abandonment of the old beds.

In addition to sedimentation attributable to changes in water flow, all forms of mariculture result in the production of organic-rich particulate wastes. Faeces (and pseudofaeces in the ease of bivalves) typically comprise the majority of the particulate wastes. Given the intensity of culture achievable, the amount of particulate material produced can be considerable. A typical oyster raft in Hiroshima Bay, for example, holds 420,000 oysters. Over a nine month culture period such a raft will produce 16 metre tons of places and pseudofaeces. With about 1000 rafts in operation In Hiroshima Bay the oyster culture will have a major impact on the sediment in the bay. The production of pseudofaeces has also been studied in connection with mussel culture in the Netherlands. For example estimated rate of sedimentation caused by mussel culture is 10mm year.

Finfish culture will not only generate faecal waste, but for those operations in which feed is provided, feed which is not ingested by the fish also contributes to the total particulate load. In salmonid culture the amount of unutilized lead can range from 1% to 30%. Faecal production for salmonids Is typically between 25 and 30% of the Ingested feed. Solid waste from salmonid farms will vary, depending on culture techniques and husbandry practices, but II has been estimated that the production of 1kg of salmon or trout will generate between 0.5 and 0.7kg of particulate waste. The fate of particulate wastes will depend upon their settling velocities and hydrodynamic conditions in the area, but generally a large proportion of the wastes, particularly uneaten food will be deposited in the immediate vicinity of the culture site. In low turbulent environments wastes accumulate typically as a soft, flocculent layer up to 40cm or more in thickness. The deposition of organic wastes from fin and shellfish culture can result in physical and chemical changes in the substrate including:

- Increases in organic carbon.
- Increased sediment oxygen consumption rates.
- Decreased sediment redox potentials.
- Generation of hydrogen Sulphide and methane.
- Increases in organic and Inorganic nitrogen content.

- Increases in phosphorus.
- Increase in silicon.
- Increase in calcium, copper and zinc.

The accumulation of organic-rich sediments and consequent physical/chemical changes in the vicinity of the culture site can have significant adverse consequences for the culturist. For example culture of shellfish in shallow embayments in Japan has led to nutrient and organic enrichment of the bottom sediments and consequent declines in culture productivity. Clam (Anadara sp.) growers in Japan have adopted a strategy of bed relation. Alter harvest a period of 1.5 years is required before the beds are reseeded. During this time the area is repeatedly trawled to accelerate remineralization.

Finfish culturists have also been confronted with fish mortality and declines in productivity associated with the accumulation of particulate waste. The accumulation of enriched sediments may be harmful to the fish either because of outgassing of hydrogen Sulphide, or the depletion of dissolved oxygen in the overlying waters. Furthermore, deterioration of water quality associated with enriched sediments may Increase susceptibility to diseases. Some Norwegian salmon farmers have found it necessary 10 disperse the accumulated sediments with propellers, or rotate cages among several sites. Thereby, allowing time for recovery of the sediment.

In addition to the effects on the cultured organisms themselves, the accumulation of organic-rich sediments has pronounced effects on the benthic macro-Invertebrates. In cases of high organic Input the sediment might become azote but such an effect is usually Iimited to the sea bed directly beneath the culture structures. At a greater distance generally less than 30m the effects are manifested by a proliferation of opportunistic species and a loss of many species Intolerant of the physical chemical consequences of organic enrichment. The spatial and temporal effects of mariculture on benthic communities are comparable to those reported from other sources of organic enrichment.

A good example of the effects of mariculture on the benthos is provided by studies of a Swedish mussel site where the deposition of places and pseudofaeces created several centimeters of sediment each year. Within six months after the start of culture, brlttle stars had disappeared and species originally dominant in the unimpacted community decreased in number and finally disappeared after 15 months. Opportunistic species became established in the culture area concurrently with the decline of the original fauna. Within six months, large populations of Capitella capitata were established, and the species later reached densities as high as 20,000 Individuals/m^2. Other opportunistic polychaetes (Scoelepis fulginosa and Microphrhslmus sczelkowil) appeared after one year of culture. Total abundance and biomass decreased initially, but then illustrated widely depending on densities of the opportunistic polychaetes.

Manson and Linden also monitored the recovery of the benthos after removal of a mussel longline that had been in production for three years. Six months after removal the bottom was still covered by 20-40 cm of mussel shells and Sulphide rich sediments. The benthos was numerically dominated by opportunistic species indicative of disturbance. Monitoring continued for a year and a half after mussel removal, during which only very Iimited macro benthic recovery was observed. In relation to salmonid culture, Gowen found a decline in macro benthic species richness and biomass three months after Initiation of culture. Following removal of the cages, sedimentary conditions showed Improvement after three months, but the macro faunal community showed evidence of substantial alteration relative to reference stations even eight months after cessation of culture.

Algaculture

Algaculture is the commercial cultivation of algae. Algae (the singular is "alga," Latin for "seaweed," but you'll rarely find just one) are simple green plants that grow in water. Their green color means they produce their own food using photosynthesis, just like grass, trees and corn. Algae come in two main forms. Macroalgae are seaweeds. Kelp grows to more than 180 feet (55 meters) long in the ocean. Nori is the variety you'll find wrapped around your sushi. Microalgae are tiny, single-celled plants that float in the water, each one visible only through a microscope.

Algaculture is nothing new. Seaweed was first cultivated in Japan at least 1,500 years ago and algae production is still a big business there. Dulse has long been eaten in the British Isles and the microalgae spirulina were harvested by the Aztecs of 16th-century Mexico. In addition to providing human food, seaweeds have been used for fertilizers. They provide the food thickener carrageen and other gelling agents and stabilizers that show up in everything from soup to toothpaste. Worldwide, algae production is a $6 billion business.

Today, algae are attracting new interest and research investment because of their potential to provide energy and combat environmental threats. Part of the organic mass of algae takes the form of oil, which can be squeezed out and converted to biodiesel fuel. Algae beat land plants hands down in the efficiency with which they produce oil. Some varieties of algae yield an oil that can be refined into gasoline and even jet fuel. The carbohydrate portion of the plants can be fermented for ethanol production.

Algae can convert waste carbon dioxide, a greenhouse gas that pours from smokestacks, to usable products. They can help clean dirty water, converting pollutants to biomass. They have additional uses in pharmaceuticals and cosmetics. With all this potential, this "weed" certainly seems to deserve a closer look.

The Promise of Algae

Why have algae generated excitement and attracted research investment in recent years? Like all green plants, algae contain chloroplasts in their cells. These tiny structures are charged with chlorophyll, a molecule that uses light energy to combine carbon and water into a simple sugar. The cells further transform some of these sugars into proteins and lipids or oil.

But if algae are doing the same thing as corn, wheat and apple trees, why bother raising them? After all, corn on the cob, sweet rolls and apple pie taste better than seaweed to most of us:

- Productivity: Algae are super fast-growing. Land plants take months or years to reach maturity. Algae can complete their entire life cycle in a single day. Some algae can double their biomass in just an hour.

- Efficiency: When it comes to converting solar energy to biomass, algae are all business. Because they're supported by and take their nutrients directly from water, they need no roots, stems or flowers. Land plants use as much as 95 percent of their energy building the structures they need for support, feeding and breeding.
- Concentration: Because of their efficiency, algae can be grown in a very concentrated space. They produce up to 100 times more oil per acre than land plants.
- Versatility: It's estimated that there are more than 70,000 species of algae, many of them not yet classified. Growers can pick ones suited to conditions and goals, selecting varieties for specific temperature range or water salinity.
- Non-competition: Algae don't compete with current crops for land or fresh water. They can be grown in ponds in locations, like deserts, that don't sustain land plants. Some varieties prefer saline or polluted water.

Attracted by all these advantages, algae cultivators have been working diligently to come up with efficient and economical ways to grow and harvest the plants. The cost factor is currently the great challenge that must be overcome to make algae commercially viable.

Commercial Cultivation of Algae

All algaculture requires three basics: water, light and nutrients. Water's the easiest. It doesn't need to be potable; different types of algae grow nicely in fresh water, salt water and dirty water. Sunlight, because it's free, is the preferred light. But sunlight reaches only 3 or 4 inches (7 to 10 centimeters) into a mass of algae, so growers must agitate the algae to expose all of it to the light. The main nutrient is carbon dioxide, which can come from the air or other source. Agitation or bubbling dissolves it into the water. The grower must supply other nutrients, like nitrogen and trace elements, if they aren't already in the water. There are three basic systems for cultivating algae, each with its advantages and disadvantages:

- Open pond: The simplest and cheapest way to grow algae is in large, shallow ponds. The water is often divided into concentric lanes or raceways, with paddlewheels to move the algae mixture in a circle. This helps bring algae to the surface, where they're exposed to light, and mixes nutrients and carbon dioxide into the liquid. The open-pond method produces less algae biomass than other methods. It loses water to evaporation, so more must be added. And it allows for contamination by predators or undesirable algae.
- Closed pond: This method is similar to an open pond, but the water is covered by a Plexiglas greenhouse. This raises the cost, but allows greater control of the process. It reduces evaporation and contamination and extends the growing season. Growers can raise algae year-round if the space is heated.
- Biophotoreactor: A completely closed system, the biophotoreactor consists of glass or acrylic tubes where the algae are exposed to light. Pumps move the water, nutrients and algae through the tubes and storage tanks. Some reactors automatically harvest the algae when they're ready. This approach gives growers the most control over the process and

the most efficient way to produce algae biomass. But it's also the most costly to set up and operate.

All of these systems are designed for growing microalgae, the one-celled varieties that float in water. Growers usually cultivate macroalgae in the open sea. The water already contains the nutrients the algae need and provides good growing conditions. The traditional method was simply to harvest wild seaweed, and this is still done in coastal areas around the world. With increased demand, growers began to cultivate seaweed. For some varieties, such as kelp, spores are attached to ropes that are then anchored in the ocean and the seaweed is allowed to grow. Other types grow from pieces of seaweed that are fixed to nets or deposited in pools.

Agriculture has been around for 10,000 years. Algaculture is relatively new. Scientists and engineers are actively studying the best ways to raise algae efficiently. The harvesting of plants is another subject of intense research.

Figure: A worker stacks baskets of harvested algae at the waterfront.

Harvesting microalgae means removing the microscopic plants from the water in which they grow and concentrating them into a paste. The grower then needs to remove the moisture, leaving a dense biomass. The minute size of microalgae presents a problem when it comes to harvesting.

One method is filtration. The grower can run the water containing the algae through a cellulose membrane whose pores are smaller than the algae cells. This can be difficult because filters quickly fill up with algae and become clogged. Researchers are looking for better ways to efficiently filter algae.

Flocculation, another method of harvest, means getting the algae to clump together. Adding chemicals or types of algae that naturally clump can cause microalgae to form clumps that become easier to gather. Another way to harvest algae is by flotation. Here, the grower uses compressed air to create a froth of bubbles and algae that brings the tiny plants to the surface where they can be skimmed off. A centrifuge is yet another harvest method. Spinning a container filled with water and algae causes the algae to collect in one end.

In order to harvest their crops most effectively, algaculture growers sometimes combine these methods. They might use flocculation to form algae clumps, then separate them with flotation or a centrifuge. Coming up with a truly efficient way to harvest microalgae is a key to bringing down the cost of cultivation.

Harvesting macroalgae involves different problems. Gathering wild seaweed is a labor-intensive process. Some types of seaweed grown in controlled conditions can be gathered in nets. Kelp raised on ropes can be hauled out and hung up to dry. Kelp forests in shallow seas can be mowed by machines, taking off the tops of undersea kelp beds.

Once harvested, algae must be drained of its water and dried. A centrifuge can spin water out, but is relatively expensive. Some systems combine harvest and processing, spreading the algae on belt filters that let the water drain through, then removing more water using a capillary medium that draws water out of the biomass of algae.

The next step is to break down the cell walls of the algae in order to extract the oil inside. The algae are put through a screw or piston press. Chemicals, electromagnetic pulses or ultrasound may also be used to break down the cells. When the oil has been drained off, the remaining biomass is compressed into a cake to be used as to supplement animal feed or as a fertilizer. Algae have found a wide range of uses, the most exciting ones in the energy field.

Uses of Algae

The buzz about algae is that it's an ideal source of renewable energy and could be the ultimate green fuel. Research by the U.S. government and companies like Boeing, Chevron and Honeywell are developing ways to make algaculture an economically viable foundation for a new generation of energy. Part of the attraction is the range of fuels into which algae can be converted:

- Biodiesel: It is the simplest way to tap algae's energy potential. Like any vegetable oil, oil from algae can be chemically transformed into biodiesel fuel. Compared to land plants like soybeans or corn, algae use less land and fresh water, grow faster and have higher concentrations of oil.
- Refined transportation fuels: These fuels are another area of promise for algae. Some algae produce oil that can be refined into gasoline or even jet fuel, and without the sulfur and nitrogen compounds in petroleum. Manufacturers can process it in the same refineries as petroleum-based stock. In 2011, the first commercial jet flight powered by algal oil flew from Houston to Chicago.
- Ethanol: It is commonly added to gasoline, can be produced from algae as well as land plants. Besides oil, algae are made up from carbohydrates and cellulose walls. These materials can be fermented by yeast into ethanol or grain alcohol.
- Methane: It is the main ingredient in natural gas, is produced when bacteria digest algae. A clean and versatile fuel, methane can be used to produce electricity or power vehicles. It represents another biofuel option for algae.

Algae actually thrive on polluted water, which means they can be used for waste water treatment.

Algae turn pollutants from municipal, industrial or agricultural waste water into usable byproducts like animal feed or biomass for conversion to energy. Algae naturally accumulate heavy metals for removal or recycling.

Because carbon dioxide, the greenhouse gas that contributes to climate change, is algae's favorite food, the plants can be used for carbon capture. They convert the gas to organic carbon compounds at a far faster clip than land plants. One pound (453.6 grams) of algae consumes 2 pounds (907.2 grams) of carbon dioxide. Feed the waste gas of a coal-burning power plant into a mass of algae, and they literally eat it up. Waste gas can be stored for permanent elimination from the atmosphere, or converted to fuel to cut the use of fossil fuels.

Algae continue play a role as human food and supplements. People eat seaweed in salads and sushi and take supplements made from the microalgae spirulina. Algae provide complete protein, omega-3 fatty acids and vitamins. Carageen is extracted from red seaweed known as Irish moss and used as a thickener.

Algae are also being used as feed for cattle and for marine animals like shrimp and shellfish. The biomass left after algae have been processed can sometimes be applied as organic fertilizer to farm fields. Algae find minor uses in cosmetics and pharmaceuticals as well.

Environmental Impact of Aquaculture

The environmental impact of aquaculture is completely dependent upon the species being farmed, the intensity of production and the location of the farm. Additionally, new strategies and technologies have emerged and have proven that it is possible to have sustainable aquaculture.

Fisheries are without doubt significant natural resources, particularly for parts of Africa and Asia. However, these areas rely heavily on fisheries for food security are places where the population is growing exponentially, meaning there are even more mouths to feed. Fish farming, when done sustainably, can be the answer to filling the gap in seafood supply that stressed fisheries are creating.

In the past, when the aquaculture industry was just getting its footing, certain factors inhibited the industry from producing fish sustainably. The intention of fish farming was never to impact the environment, but to increase food security. However, environmental problems did arise. There was no shortage of negative press, and these stories have stayed with the public. Common criticisms were related to nutrient and effluent build-ups, the impact of fish farms on local wild fisheries with respect to disease and escaping, and environmental degradation due to the site's location.

The negative environmental impacts aquaculture has had are nuanced. Nutrient buildup happens when there is a high density of fish in one area. Fish produce waste, and their waste has the potential to build up in the surrounding area. This can deplete the water of oxygen, creating algal blooms and dead zones. Farmers' usage of antibiotics to prevent disease created concern about the effect of the drugs on the ecosystem around the cages, including wild fish. Many also worried that the escape of nonnative fish would cause wild fish to compete for food, potentially displacing the native fish.

These were all valid criticisms given that the industry was just beginning to learn how to cope with issues as they came up, as any new industry does.

Fortunately for the aquaculture industry and for the planet's well-being, much progress has been made in the name of sustainability.

Regulatory agencies have recognized the effects of nutrient and effluent buildup with implementing measures to prevent this from happening. Siting the location of the fish farm in an area with strong currents to disperse the effluent as well as moving the farms from time to time to prevent impacting one specific area more than others is some ways the industry is combatting this environmental issue. Additionally, land-based aquaculture has a minimal impact on local ecosystems.

Antibiotic use is declining and safe and effective vaccinations have since been developed for farmed fish and are now widely used. As for the issue of escaped fish from farms, which does still happen occasionally, underwater cameras now closely monitor cages, and divers regularly inspect cages with this in mind.

Positive Environmental Impacts

Farmed seafood is incredibly resource efficient, especially when compared with other animal proteins (beef, pork, chicken). The feed conversion ratio, which is the measurement of how much feed it takes to produce the protein, is 1:1. This means that essentially one pound of feed produces one pound of the protein. Beef, pork and chicken's feed conversion ratios vary between 2.2- 10. As a result, seafood's protein retention, as well as energy retention is remarkably high as well.

As farmed fish are closely monitored in comparison to wild fish, farmers have more control over variables. This can positively impact the environment and the fish. Farmed fish are generally free of environmental contaminants like mercury and heavy metals, as they exclusively eat human-processed feed. Fish feed's toxin levels are regulated.

The farming of filter-feeders, like shellfish, can improve water quality. These creatures eat excessive nutrients in the water, which, in turn, prevent the buildup of effluent. Filter-feeders are often integrated into the farming of other species, like finfish, to use uneaten feed and fish waste as food, offsetting the farm's environmental impact. This system is called polyculture, or integrated multi-trophic aquaculture (IMTA).

Sustainable fish feeds are on the rise. Fishmeal and fish oil used in feeds may come from trimmings from processing plants. Additionally, substituting plant proteins for fishmeal in feeds is also becoming more prevalent.

Pond Culture

Pond culture is the most common method of fish culture. Water is maintained in an enclosed area by artificially constructed ponds where the aquatic animals such the finfish and shellfish are reared. The ponds may be filled with canal water, rain water, bore well water or from other water sources. The pond must be constructed after proper site selection. The climate, topography, water availability and soil quality of the region influence the character of the fish pond. Based on these factors, ponds are primarily of two types, namely:

- Watershed pond.
- Levee Pond.

Watershed Pond Systems

For watershed ponds, water required to fill and maintain the pond water is entirely sourced from the watershed runoff, though groundwater (bore well) and surface water (stream and reservoirs) can be used as additional water sources. The major factors to be considered are soil type, characteristics, topography and water supplies:

- The soil type influences how well the ponds will hold water; mud and mud-silt type soils are preferred for pond construction because it prevents leakage. The good quality soil containing a lower limit of 20 percent clay is necessary for making ponds.
- The topography determines the size and shape of the ponds. The watershed ponds should not be more than 10 feet deep. The size of the ponds should be less than 20 acres for better management. Sites, where huge individual ponds could be built, can be divided into smaller ponds built in series.
- The availability and quality of water determine where and what type of pond should be made. Growing and harvesting are more challenging in watershed ponds than levee ponds due to erratic water supplies, uneven bottom and side, and size and excessive depth of the dugout.
- Advantages of this pond system include Construction of Pond is inexpensive, free water is available and there is less competition for water from other agriculture activities.

Levee Pond Systems

Levee ponds are created in flat land areas where there is inadequate water to fill the ponds from the watershed. The groundwater is typically used to fill the ponds.

Prior to building of ponds make sure whether the site is suitable for pond construction by looking at the characteristics of soil type, quality and ground water availability. The type of soil influences the pond productivity and life expectancy. It is necessary to have 20 percent of clay in the soil to prevent the leakage of the pool. The pond topography decides the size of the pond; generally, rectangular pond size is chosen due to its greater simplicity and ease of harvesting and feeding. Curved and irregular ponds are not recommended as it is hard to manage water quality. The water quality should also be considered before construction of the pond. Levee Pond is more suitable for growing and harvesting fish than the watershed pond.

Culture Species

Commonly raised species in freshwater ponds are the carps, tilapia, catfish, snakehead, eel, trout, goldfish, gouramy, trout, pike, tench, salmonids, palaemonids, and the giant freshwater prawn Macrobrachium. In brackish water ponds, common species include milkfish (Chanos chanos), mullet (Mugil sp.) and the different penaeid shrimps (Penaeus monodon, P. orientalis, P. merguiensis, P. penicillatus, P. semisulcatus, P. japonicus, and M. ensis). The more popular species for culture in marine ponds are the sea bass, grouper, red sea bream, yellowtail, rabbit fish, and marine shrimps.

In Asia, where the bulk of world production from aquaculture emanates, fish ponds are mostly freshwater or brackishwater, and rarely marine. In China and most of the Indian sub-continent, pond culture is traditionally dominated by freshwater species, mainly the carps, usually in polyculture and/or integrated with animal husbandry. In Southeast Asia, fish ponds are predominantly brackishwater, with milkfish and penaeid shrimps grown either in polyculture or in monoculture.

Recently in Latin America and the Caribbean, brackishwater pond culture of penaeid shrimps has expanded rapidly, as it has in some parts of Asia.

In Africa, the tilapias and carps dominate aquaculture production. Controlled breeding is also carried out in ponds with goldfish, trout, Bagrus and, to a lesser extent, Lates niloticus, Heterotis niloticus, and Clarias lazera. Ten species of molluscs belonging to four genera (Crassostrea, Mytilus, Venerupis and Pinctada) are cultured. Crustacean culture has yet to be developed on a significant scale.

Site Selection

Proper site selection is recognized as the first step guaranteeing the eventual success of any aquaculture project and forms the basis for the design, layout, and management of the project. For fish ponds, especially those to be used for coastal/brackishwater aquaculture of high-value species like shrimps, site selection is critical and should be given utmost attention.

The following guidelines for the selection of a suitable site for coastal fish ponds are:

- Soil Quality: Preferably, clay-loam, or sandy-clay for water retention and suitability for diking; alkaline pH (7 and above) to prevent problems that result from acid-sulphate soils (e.g., poor fertilizer response; low natural food production and slow growth of culture species; probable fish kills).

- Land elevation and tidal characteristics: Preferably with average elevation that can be watered by ordinary high tides and drained by ordinary low tides; tidal fluctuation preferably moderate at 2-3 m. (Sites where tidal fluctuation is large, say 4 m, and are not suitable because they would require very large, expensive dikes to prevent flooding during high tide. On the other hand, areas with slight tidal fluctuation, say 1 m or less, could not be drained or filled properly).

- Vegetation: Preferably without big tree stumps and thick vegetation which entail large expense for clearing; areas near river banks and those at coastal shores exposed to wave action require a buffer zone with substantial growths of mangrove. (The presence of Avicennia indicates productive soil; nipa and trees with high tannin content indicate low pH).

- Water supply and quality: With steady supply of both fresh and brackish water in adequate quantities throughout the year; water supply should be pollution-free and with a pH of 7.8-8.5.

- Accessibility: Preferably readily accessible by land/water transport; close to sources of inputs such as fry, feeds, fertilizers, and markets, fish ports, processing plants, and ice plants; and linked by communication facilities to major centers.

Pond Layout

The layout of the pond system depends on the species for culture and on the size and shape of the area, which in turn determines the number and sizes of ponds and the position of the water canals and gates. A fish farm is considered properly planned if all the water control structures, canals, and the different pond compartments mutually complement each other. A complete fish farm has nursery and grow-out ponds and, in some instances, transition ponds for intermediate-sized fish/ shrimp, all of which are properly proportioned and positioned within.

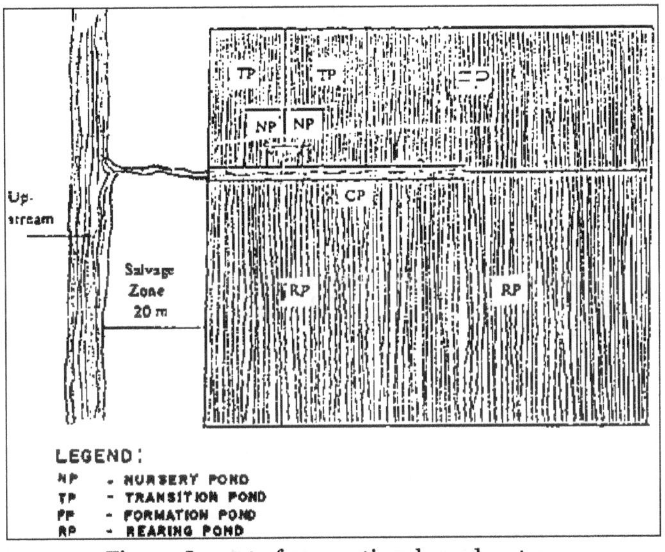

Figure: Layout of conventional pond system.

Milkfish culture in brackishwater ponds in the Philippines follows the traditional practice of providing for nursery, transition, and rearing operations. In some cases, formation ponds are used for additional growth or stunting of fingerlings prior to stocking in rearing ponds. The nursery ponds comprise about 1-4% of the total production area while the transition and formation ponds constitute about 6-9% of total area.

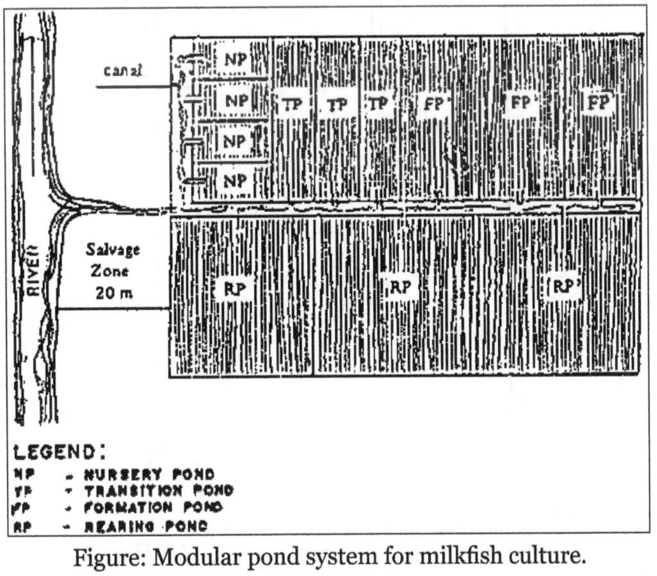

Figure: Modular pond system for milkfish culture.

It has been suggested that a similar progressive culture scheme be adopted for shrimp pond culture when no supplementary feeding is practiced. For growing to a medium size, a two-stage progression composed of a nursery pond (NP) and a rearing pond (RP) is adequate (Fig. 3); for growing to larger sizes, a three-stage progression composed of nursery, transition, and rearing ponds is recommended.

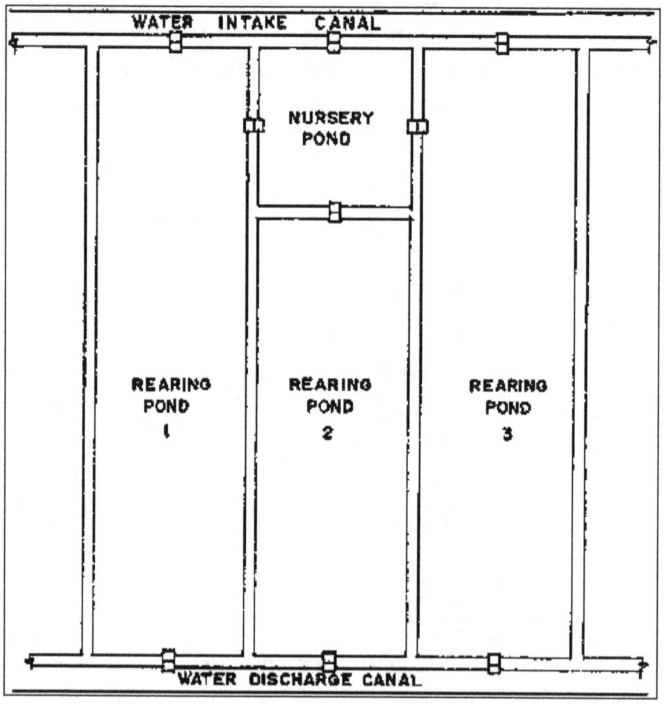

Figure: Pond layout with one nursery pond and three rearing ponds.

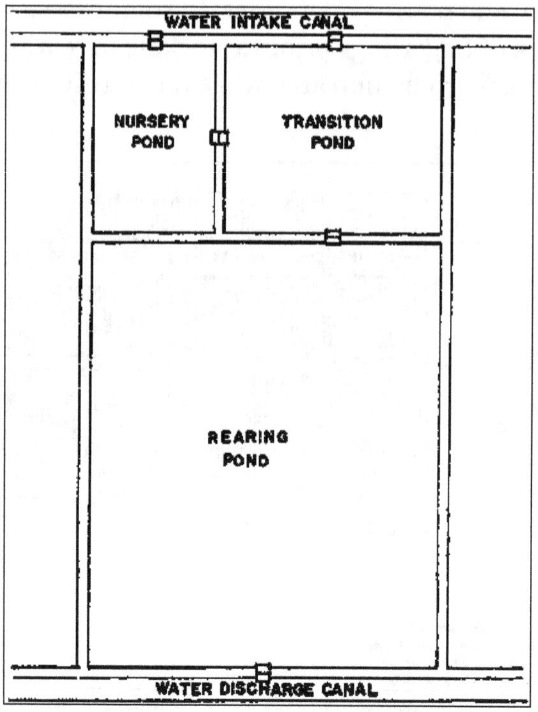

Figure: Pond layout with one nursery pond, one transition pond, and one rearing pond.

In general, however, shrimp monoculture uses direct stocking of post larvae in rearing ponds and therefore requires only one type of pond with separate inlets and outlets for better circulation and aeration.

Design of Pond Facilities

A fish pond system consists of the following basic components:

- Pond compartments enclosed by dikes.
- Canals for supply and drainage of water to and from the pond compartments.
- Gates or water control structures to regulate entry and exit of water into and from the pond compartments.

Pond Compartments

They are usually rectangular in shape although in Indonesia, running water ponds are generally triangular, raceway-shaped, or oval. They vary in size from less than a hectare to several hectares each; sometimes up to 20-50 ha in size. However, with the new intensive methods, the trend is to use smaller units for flexibility and ease of management.

The elevation of the rearing pond bottom for milkfish is usually such that only a maximum of 40 cm of water can be held in the ponds during the culture period. For new shrimp ponds, the minimum water depth is 1 m.

The entire pond system is enclosed by a perimeter dike and the individual pond compartments are separated from each other by partition dikes. The outer perimeter dike is usually wider and higher than the inner partition dikes and serves to protect the entire fish pond area from flooding and destruction brought about by tide and wave action. The inner dikes are narrower and shorter.

The design of the dikes depends primarily on soil characteristics. Dikes are usually earthen although intensive shrimp ponds are concrete-lined or brick-lined as in Taiwan (PC). The side slopes are designed for structural stability, the ratio of horizontal length to height ranging from 1:1 to 1:3. The height and width of dikes depend on the type (primary, secondary, or tertiary), tide conditions, flood level, pond water depth, soil shrinkage, and freeboard.

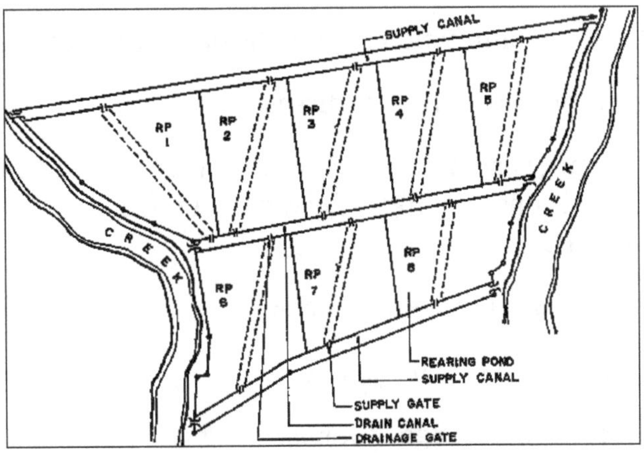

Figure: (A) Typical cross sections of dikes.

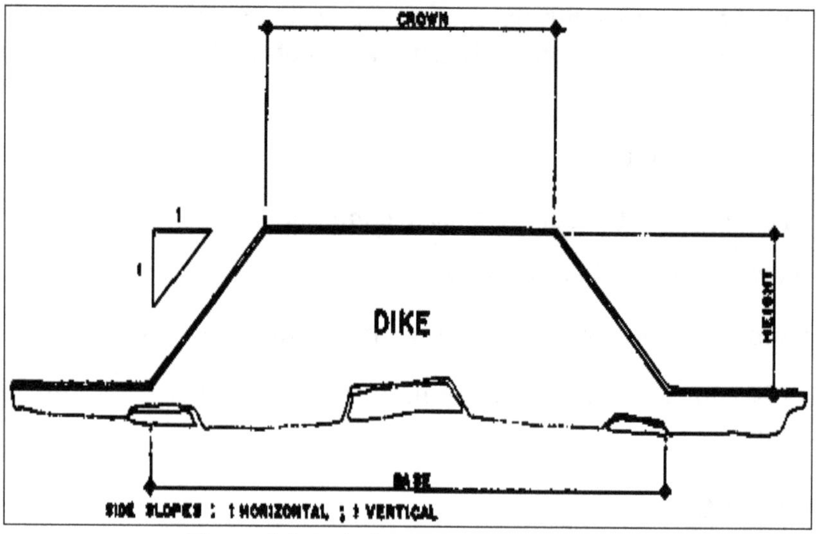

Figure: (B) Typical cross sections of Dikes.

The following slopes are recommended for dikes built with good clay soil:

- 2:1 when dike height is above 4.26 m and exposed to wave action.
- 1:1 when dike height is less than 4.26 m and tidal range is greater than 1 m.
- 1:2 when tidal range is 1 m or less, and dike height is less than 1m.

The dike crown should not be less than 0.5 m and the main dike surrounding the farm should be 0.5 m above the highest dike or flood level recorded in the locality.

Water conveyance structures: (canals/channels) supply new water into the pond and drain out old water. They also provide the facility for holding and harvesting of fish and of serving as waterways for transporting farm supplies. Traditional milkfish ponds usually have only one canal that is used for both supply and drainage. Shrimp ponds have separate supply and drainage canals. Canals which are to be used for harvesting should be 30 cm below the level of the pond bottom to allow draining of pond water.

Having separate water intake and discharge canals in a pond complex brings about the following advantages:

- Better filling and non-contamination of pond by discharge from other ponds.
- Greatly reduced possibility of spread of disease.
- Maintenance of constant head in intake canal thus reducing water loss through leaks/seepages in pond dikes and consequently reducing leaching of acids into the ponds from dikes with acid-sulphate soils.
- Absence of conflict of usage between farmers.
- Better water exchange for individual ponds.
- Possibility of effecting flow-through systems.

The width of the canals depends on the amount of water they must carry. The following should be taken into account when designing canals:

- Volume of water to be held in the ponds.
- Time requirement for filling or draining the pond.
- Amount of rainfall which must be carried off in a given period of time.
- Elevation of canal bottom in relation to tide.
- Other uses like transportation, harvesting of milkfish, and holding of broodstock.

Diversion canals are constructed where there is much runoff from adjoining areas, to prevent sudden salinity changes and the possible entry of polluted, pesticide-loaded water and/or of silted water into the pond complex.

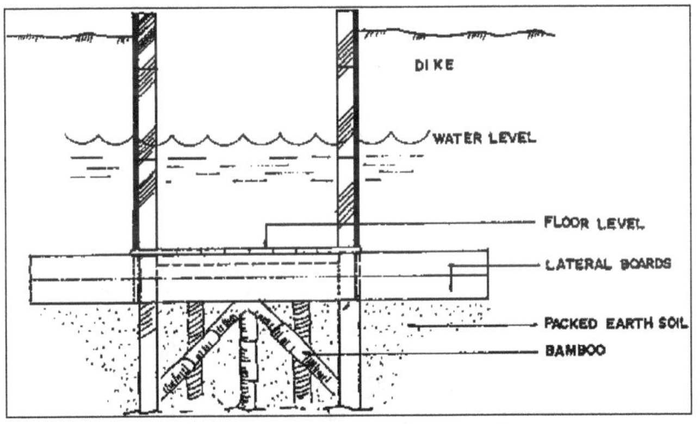

Figure: Use of culvert pipes as secondary gates.

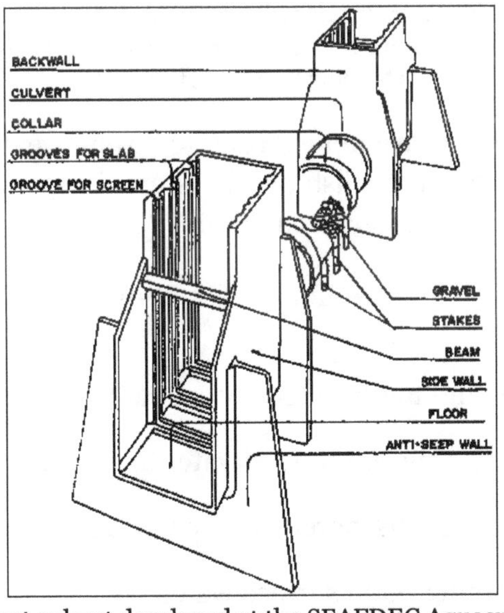

Figure: Ferro cement culvert developed at the SEAFDEC Aquaculture Department.

The entry and exit of water into ponds through the canals is regulated or controlled by gates. Main gates regulate the exchange of water between the pond system and the tidal stream or sea, and may be constructed of reinforced concrete or wood. Reinforced concrete is more expensive but lasts longer. Such a gate has one or multiple (2, 3, 4, etc.) openings depending on the relative size of the pond unit to be served. A recent innovation for a smaller and less expensive main gate is the monk-type gate which uses culverts usually made of concrete hollow blocks. The SEAFDEC Aquaculture Department has also introduced the open sluice gate made of Ferro-cement.

Secondary gates, which regulate water exchange between the ponds and the canals, are usually made of wood. Pipes or culverts can also be used for smaller ponds such as nursery or fry ponds and transition ponds for milkfish culture. Secondary gates are now usually located toward one end of the narrower side of the pond compartment to give good turbulence and circulation during the filling and draining.

Shrimp ponds are provided with separate supply and drainage gates to effect flow-through water management and facilitate water exchange through supply and drainage canals. Inlet and outlet gates are best located at opposite corners of the same pond, across which a diagonal trench, about 5-10 m wide and 0.3-0.5 m deep, extending from inlet to outlet gates is recommended for convenient draining of water.

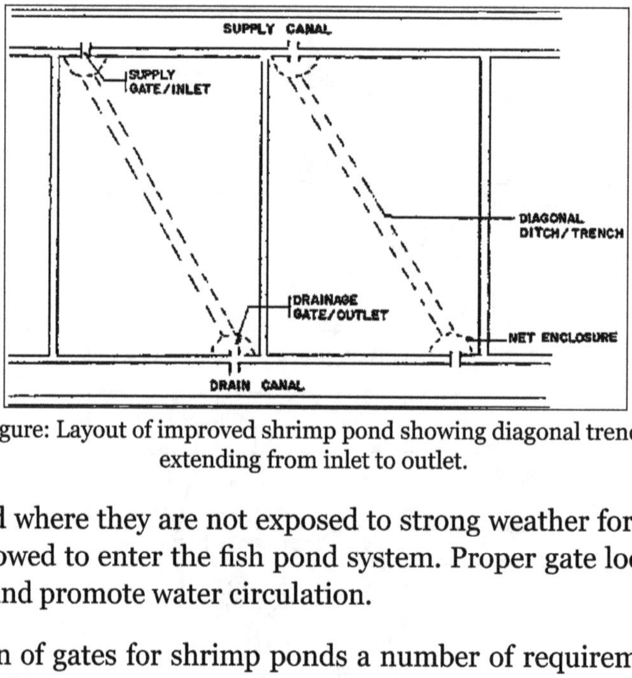

Figure: Layout of improved shrimp pond showing diagonal trench extending from inlet to outlet.

Gates should be located where they are not exposed to strong weather forces and where water of good quality can be allowed to enter the fish pond system. Proper gate location can also serve to aerate the pond water and promote water circulation.

During the construction of gates for shrimp ponds a number of requirements should be kept in mind and the gates should:

- Be durable, water-tight, and made of locally available materials.
- Have adequate capacity for the amount of water to be taken in or drained.
- Allow water to be taken in or discharged at the bottom.
- Have provisions for draining pond surface water.
- Have gate bottom elevation that permits complete draining of pond water.
- Have slots or grooves for the placement of outside and inside screens to prevent undesirable species from entering the pond and the shrimps from leaving the pond.

- Have place for net installation for harvesting.
- Be easy to operate.

Pond Management

Pond management techniques for finfish and shrimp culture, while varying slightly depending on the specific biological requirements of the culture organism, the type of culture system, and the culture environment (freshwater, brackish water, and marine), are similar in that they involve the following basic activities:

- Pond preparation/conditioning,
- Stocking,
- Feeding and/or fertilization (depending on the culture system used),
- Water management,
- Pond maintenance,
- Harvesting.

Variations would consist mainly of differences in application rates of fertilizers, lime, pesticides, and feeds; stocking rates and sizes of stocking material; rate of water change; and harvesting techniques.

Extensively managed systems generally require the least management, with no supplemental feeding and minimal water exchange on account of the low stocking density used. On the other hand, intensively managed ponds require full artificial feeding and substantial water management to ensure optimum culture conditions for the species being reared.

Pond Preparation

Ponds are totally drained and the pond bottoms dried prior to the application of pesticides. Tobacco dust, derris root/rotenone powder, teaseed cake/powder, or Gusathion -A are used to eliminate predators and/or wild species that may eventually compete with the cultured organisms for food and space. Teaseed cake is perhaps the best fish poison to use in brackish water ponds to selectively kill unwanted fish without damaging the shrimps and without affecting rotifers and copepods which are feed for shrimps. On the other hand, rotenone is most effective in fresh water and works better in low-salinity water.

Ponds with acid-sulphate soils are repeatedly dried and flushed, i.e., filled and drained to remove the acids formed by pyrite oxidation. Agricultural lime is then applied to correct soil pH and bring it up to at least 6.5. Brackish water ponds are usually treated by spreading 1.5 t of agricultural lime per ha, followed by another 1.5 t worked into the soil.

To stimulate and maintain the growth of natural plankton, organic (e.g., chicken manure) or inorganic fertilizer (e.g., urea, ammonium phosphate) are applied to the pond bottom. After fertilizer application, water is let in to a depth of about 20-40 cm and gradually increased to 1 m a week after fertilization. Intensively managed ponds or ponds where artificial feeding shall be given, do not need to be fertilized. Extensive ponds need regular fertilization during the culture period to maintain the growth of natural food. Semi-intensive ponds may use a mix of fertilization and supplementary feeding.

Table: Variations in pond management techniques commonly used for different species.

Species	Stocking Rate	Fertilization	Feed Type	Rate of Water Change	Pesticides/Predator Control Type	Application rate
MILKFISH (Chanos chanos)	2 000-5 000/ha	16-20-0 at 50 kg/ha; 45-0-0 at 15 kg/ha; chicken manure at 0.5 t/ha twice weekly.	Rice bran and trash fish as supplemental feed.	Once every two weeks at high tide.	Lime; ammonium sulfate.	1 t/ha 10 g/m^2
TILAPIA (O. niloticus; O. mossambicus)	5 000-20 000/ha	Chicken manure at 500 kg/ha; Inorganic fertilizers at 50 kg/ha.	Rice bran, fish meal, ipil-ipil leaf meal.			
CATFISH (Clarias botrachus and monocephalus)	60-300/m2		9 parts trash fish and 1 part rice by-products.	When necessary.		
PENAEIDS	From as low as 15 000 to as high as 300 000/ha	Chicken manure at 1-2 t/ha followed by inorganic fertilizer at 75-150 kg/ha mono-ammonium phosphate (16-20-0) and 25-50 kg/ha of urea (46-0-0).	Supplemental feed of rice bran with trash fish, mussels, and clam meat; artificial/formulated diets with 40% CP.	20-30% once every week or every two weeks for low density ponds; 5-20% daily for semi-intensive to intensive ponds.		

Stocking

After the pond is prepared, fish fingerlings or shrimp post larvae are stocked at the appropriate density depending on the culture strategy, size of pond, and the size of fingerlings, among others. The fingerlings are properly acclimated and conditioned prior to stocking and weak or diseased fish eliminated. Stocking is usually done in the early morning or late afternoon.

Feeding

Fish/shrimp grown in semi-intensive and intensive culture ponds are given supplementary and full artificial feeds, respectively, the former to augment the natural food in the pond, the latter to totally replace the natural organisms in the water as a source of nutrition.

A wide variety of feed ingredients is used to prepare supplemental/artificial feeds. The simplest fish feeds are prepared at the pond site using locally available raw materials like rice or corn bran, copra meal, and rice mill sweepings as sources of carbohydrates. These are usually mixed with animal protein like trash fish/fish meal, shrimp heads, and snail meat. Supplemental feeds for tilapia are prepared using 80% rice bran and 20% fish meal. Those for shrimps in improved extensive culture (low-density stocking but given dietary supplements for increased growth/production) usually include fresh raw materials like snail/mussel/clam meat or carabao hide and other slaughterhouse leftovers.

Commercial feed preparations are also available now in a wide range of brand names, mostly for semi-intensive and intensive shrimp culture. (Taiwan (PC), Japan, and the USA are the top producers of commercial fish/shrimp feeds.) These commercial diets consist of a number of ingredients like fish meal, blood meal, bone meat, and shrimp head meal (to serve as attractant for the shrimp), together with vitamin and mineral premix and carbohydrate sources like rice/corn bran or wheat. The crude protein (CP) content of these shrimp feeds is generally not lower than 30% to satisfy the high animal protein requirement of shrimps, actually estimated to be about 40% during the earlier stages of growth.

Commercial feeds usually come in various formulations to match the protein requirement of the culture organism, which as a rule, decreases with age. Thus, fish/shrimp feeds come in different forms as starter, grower, and finisher, with starter feeds having the highest CP content of about 40% and finisher feeds having the lowest CP content of about 20%. Starter feeds are usually given on the first month of culture, finisher feeds on the last month, and grower feeds in between.

Some shrimp culturists prefer not to give artificial feeds during the first two weeks of culture when the newly stocked post larvae can subsist on the plankton available in the water.

The feeding rate is computed as a percentage of the estimated animal biomass in the pond, with higher rations given when the animals are small and gradually decreasing as they become bigger. The daily feeding rate usually starts at 5% and 10-15% of estimated biomass of fish and shrimps, respectively, and decreases to a low of 2% and 5%, for fish and shrimps, respectively, toward harvest.

The daily feed rations are given in equal portions during the course of a day. Freshwater fish like tilapia are usually fed twice a day - early morning and late afternoon. Penaeid shrimps are fed more frequently, from three to four to as often as six to seven times a day.

Feeds are broadcast into the water and/or supplied on feeding trays. In semi-intensive and intensive shrimp ponds, small feeding boats are used by caretakers who go around the pond distributing the feed by broadcasting. At certain points along the periphery of the pond, feeding trays are submerged into the water after known quantities of feed are put on the surface, to supply feed to the shrimps in the pond as well as to monitor feed consumption and shrimp growth. The feeding tray is lifted two to three hours after the feed was supplied to check how much of it has been consumed and to see if the shrimps are healthy and feeding. Empty feeding trays may indicate that the quantity given is inadequate and may have to be increased. Conversely, full or slightly touched trays indicate excessive feed quantities and/or sluggish shrimps. The feeding ration is subsequently adjusted accordingly to optimize feed utilization.

By monitoring the feeding tray, one can get a good indication of the sizes and quantity of shrimps present in the pond without a need for cast-netting or actual sampling, since shrimps are invariably found on the tray when it is lifted out of the water.

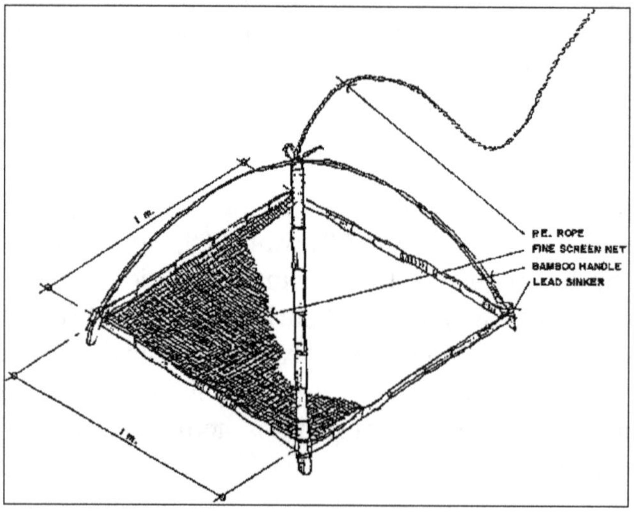

Figure: Feeding tray.

Water Management

Water in the pond is kept at certain levels for optimal fish growth. In general, a pond water depth of 1 meter is considered best for culture of tilapia, carps, and shrimps; traditional milkfish ponds can do with just 40-60 cm of water.

Pond water is not just maintained at a certain depth; its quality must also be kept high to ensure optimal growth of the culture organism. This is particularly important in semi-intensive and intensive culture systems where large amounts of metabolites are continuously excreted into the pond and where excess, unconsumed feeds add to the bottom load and serve to pollute the water.

To prevent the deterioration of the pond environment, pond water is continuously freshened by the entry of new water from the river or water source (through the supply canal) while old water is drained through the outlet/drainage gate and through the drainage canal into the sea or river.

A flow-through system of water management that allows the simultaneous entry and exit of water into and out of the pond is essential in any high-density culture system. This is affected by the provision of separate inlets and outlets for all the ponds, each inlet regulating the flow of water from the supply canal to the pond and each outlet controlling the discharge of water out of the pond into the drainage canal. Both the supply and drain gates are so designed as to bring water into and drain water out of the lower levels of the pond, where water quality tends to get poorer faster as a result of the accumulation of wastes and their subsequent decomposition.

The regular replenishment of pond water, independent of natural tidal fluctuations, is made possible by the use of pumps which draw water from the source even at low tide. Although there is no hard-and-fast rule as to the rate of water change necessary for medium- to high density aquaculture, semi-intensive culture systems usually change water at the rate of 10% daily for an equivalent total replacement of water every ten days or three times per month. Intensively managed ponds require greater water exchange in view of the much higher organic load on the pond bottom, especially toward the latter part of the culture cycle when the animals excrete more wastes.

Intensive ponds/tanks usually need to provide for aeration facilities/equipment to prevent anoxia that may lead to mass mortalities. Oxygen depletion in high-density ponds results not only from the faster rate of utilization of dissolved oxygen for respiratory activities; it is also caused by the fast rate of decomposition at the pond bottom by aerobic or oxygen-consuming micro-organisms.

Paddlewheels or other types of aerators are thus provided in the ponds to affect the infusion/introduction of greater quantities of oxygen into the water and prevent fish/shrimp mortalities. The aerators are usually operated at regular/periodic intervals for certain fixed durations during the day but especially in the early morning hours when the concentration of dissolved oxygen is known to be lowest (as a result of the absence of photosynthetic, oxygen-producing activity in the pond). Toward the end of the culture period when oxygen demand is highest, aeration may have to be provided continuously and not just sporadically as could be done during the initial stages of rearing. At that time too, water pumps usually need to be run for longer periods to affect greater water exchange.

Pond water is also regularly sampled and measurements taken of basic/essential parameters, particularly dissolved oxygen, pH, and salinity. This is important for the purpose of determining the need for corrective/remedial action to bring water quality to optimum levels and obtain good yields.

Dissolved oxygen levels are kept, as much as possible, above 5 ppm by pumping and aeration. Problems of acidity are corrected by liming. Salinity is an important parameter for penaeid culture and has to be maintained within a range of 15-25 ppt for best results. During summer months, high-salinity water can be diluted by mixing with fresh water from springs or deep wells.

Pond Maintenance

Fertilization

Aside from feeds and water management, the following pond maintenance procedures are carried out: regular application of fertilizers, lime, and pesticides; prevention of entry of predators; monitoring of the stock for growth rate determination as a basis of feeds and water management; and regular pond upkeep and maintenance.

Extensive ponds are fertilized regularly using either organic fertilizers like chicken, cow, or pig manure, or inorganic fertilizers like urea, ammonium phosphate, or both, to maintain the plankton population in the pond. The fertilizers are either broadcast over the pond water surface or kept in sacks suspended from poles staked at certain portions along the pond periphery. Semi-intensive and intensive culture systems do not require fertilization since they are not natural food-based, except for those which grow plankton-feeders like milkfish whose diet is largely algae dependent.

Liming

In addition to fertilization, ponds also need to be given regular doses of lime to maintain water pH at alkaline or near-alkaline levels (preferably not lower than six). Agricultural lime is broadcast over the pond and applied on the sides of the dikes to correct soil and water acidity.

Elimination of Pests and Predators

- Unwanted and predatory species which may have survived the application of pesticides during pond preparation or which were able to enter the pond through the gate screens or through cracks in the dikes, are eliminated by the application of pesticides, preferably organic, into the pond.
- Crabs, which are a serious problem in shrimp ponds because they are carnivorous and cause damage to the pond dikes, are not usually affected by known pesticides and are therefore best eliminated by the use of crab traps situated in the pond.
- It is also important that the gates are properly screened and the screens kept whole, to prevent the entry of small unwanted fish into the pond. Double screens are usually installed at the main intake to ensure that pests and predators are prevented from entering the pond system.

Stock Monitoring

The culture organisms are monitored closely and regularly to determine their rate of growth and the general condition of the stock. They are regularly sampled for length-weight measurements as a basis for determining/estimating their biomass in the pond and therefore their daily feed rations, as well as for making projections on harvest schedules and procurement of pond inputs.

In the first few months of culture, the feeding tray is a good tool for stock monitoring. As the organisms grow in size, cast-netting is used as a sampling tool, with those caught in the throw of the cast net providing an indication as to sizes and weights of stock. Based on the sampled weights and the daily feed consumption, it is possible to predict the available biomass (i.e., stock surviving after initial mortalities) and make projections on volume of harvest. For this purpose, it is essential that accurate records are kept for analysis at a later time. Data on initial size/weight and number of fry/post larvae stocked, average body weight at each sampling, and feed consumption on a daily basis, are important to have on file.

Regular Upkeep and Maintenance of Facilities

The pond dike and gates are checked regularly for cracks that could lead to seepages and losses of stock. The dikes are best planted with grass or vegetative cover to prevent erosion. The gates and other support infrastructure are properly maintained for efficient operation.

Harvesting

Marketable-size fish/shrimps are harvested at the end of the culture period by draining the pond and using harvesting nets to catch the fish or shrimps. The latter are harvested with a bag-net attached to the sluice gate as water is drained out of the pond at low tide. Tilapia is harvested using seine nets after the pond water is drained to half-level the night before.

Harvest of milkfish takes advantage of their behaviour of swimming against the current. The method, known in the Philippines as "pasulang" or "pasubang" involves draining 85-90% of the pond water during low tide and allowing in the water at the incoming high tide so that the fish swim against the current through the tertiary gate and into the catching pond, whose gate is closed once a large number of fish is impounded. The fish in the catching pond are then harvested by seining and the rest hand-picked.

Recirculating Aquaculture Systems

Recirculation aquaculture systems (RAS) represent a new and unique way to farm fish. Instead of the traditional method of growing fish outdoors in open ponds and raceways, this system rears fish at high densities, in indoor tanks with a "controlled" environment. Recirculating systems filter and clean the water for recycling back through fish culture tanks.

New water is added to the tanks only to make up for splash out and evaporation and for that used to flush out waste materials. In contrast, many raceway systems used to grow trout are termed "open" or "flow through" systems because all the water makes only one pass through the tank and then is discarded.

Fish grown in RAS must be supplied with all the conditions necessary to remain healthy and grow. They need a continuous supply of clean water at a temperature and dissolved oxygen content that is optimum for growth. A filtering (bio filter) system is necessary to purify the water and remove or detoxify harmful waste products and uneaten feed. The fish must be fed a nutritionally-complete feed on a daily basis to encourage fast growth and high survival.

RAS have been in existence, in one form or another, since the mid 1950's. However, only in the past few years has their potential to grow fish on a commercial-scale been realized. New water quality technology, testing and monitoring instrumentation, and computer enhanced system design programs, much of it developed for the waste water treatment industry, have been incorporated and have revolutionized our ability to grow fish in tank culture.

Nevertheless, despite its apparent potential, RAS should be considered a high-risk, experimental form of agriculture at this time. They can be used to culture high-densities (over a million pounds) of fish annually, but their ability to do so economically remain to be demonstrated, conclusively and repeatedly.

Benefits of RAS

RAS offer fish producers a variety of important advantages over open pond culture. These include a method to maximize production on a limited supply of water and land, nearly complete environmental control to maximize fish growth year-round, the flexibility to locate production facilities near large markets, complete and convenient harvesting, and quick and effective disease control.

RAS can be of various sizes ranging from large-scale production systems (over 1 million pounds per year) to intermediate-sized systems (500,000 pounds per year), to small systems (50,000 pounds per year). They can be used as grow-out systems to produce food fish or as hatcheries to produce eggs and fingerling sport fish for stocking and ornamental fish for home aquariums.

Intensive Production

RAS applies to the broiler house or swine barn concept, so prevalent and effectively used in modern poultry and pork production systems, to rear large numbers of fish in a relatively small space. Indoor fish farming in tanks may revolutionize fish production in the same way that confinement systems altered the pork and poultry farming industries. This is an excellent alternative to open pond culture where low densities (extensive culture) of fish are reared free in large ponds and are subject to losses from diseases, parasites, predation, pollutants, stress, and seasonally suboptimal growing conditions.

Water and Land Conserved

RAS conserve both water and land. They maximize production in a relatively small area of land and use a relatively small volume of water. For example, using a RAS it is possible to product over 100,000 pounds of fish in a 5,000 square-foot building, whereas 20 acres of outdoor ponds would be necessary to produce an equal amount of fish with traditional open pond culture.

Similarly, since water is reused, the water volume requirements in RAS are only about 20% of what conventional open pond culture demands. They offer a promising solution to water use conflicts, water quality, and waste disposal. These concerns will continue to intensify in the future as water demand for a variety of uses escalates.

Location Flexibility

RAS are particularly useful in areas where land and water are expensive and not readily available. They require relatively small amounts of land and water. They are most suitable in northern areas where a cold or cool climate can slow fish growth in outdoor systems and prevent year-round production. RAS provide growers who are geographically disadvantaged because of a relatively short growing season (less than 200 days) or extremely dry (desert) conditions, a competitive, profitable, year-round fish production system.

They can be located close to large markets (urban areas) and thereby reduce hauling distances and transportation costs. RAS can use municipal water supplies (de-chlorination is necessary) and discharge waste into sanitary sewer systems. Nearly all species of food fish and sport fish that are commonly reared in ponds including catfish, tout, and striped bass can readily be grown in high densities when confined in tank systems.

Species and Harvest Flexibility

RAS are currently being used to grow catfish, striped bass, tilapia, crawfish, blue crabs, oysters, mussels, and aquarium pets. Indoor fish culture systems offer considerable flexibility to grow a wide diversity of fish species, rear a number of different species simultaneously in the same tank (polyculture) or different tanks (monoculture), raise a variety of different sizes of one or several species to another depending on market demand and price.

RAS afford growers the opportunity to manipulate production to meet demand throughout the year and to harvest at the most profitable times during the year. This flexibility in the selection of species and harvest time allows the grower to rapidly respond to a changing marketplace in order to maximize production and profitability. RAS permit the grower to competitively respond to market price and demand fluctuations by altering harvest rates and times and the species cultured. Tank culture systems are now being used to hold and purge (depurate) contaminated of off-flavor, pond reared catfish until they are acceptable for marketing.

RAS do have some disadvantages when compared to open pond culture. They are relatively expensive systems to develop (building, tanks, plumbing, bio filters) and to operate (pumping, aerating, heating, lighting). Moreover, they are complex systems and require skilled technical assistance to manage successfully.

Constant supervision and skilled technical support are required to manage and maintain the relatively complex circulation, aeration, and biofilter systems, and to conduct water quality analysis. The danger of mechanical or electrical power failure and resulting fish loss is always a major concern when rearing fish in high densities in small water volumes.

Operating at or near maximum carrying capacity requires fail-safes in the form of emergency alarms and backup power and pump systems. The business and biological risk factors are correspondingly high. Continuous vigilance and quick reaction times (15 minutes or less) are needed to avert total mortality. However, the higher risk factor, capital investment, and operating costs can be offset by continuous production, reduced stress, improved growth, and production of a superior product in the RAS.

The RAS Design

The functional parts of a RAS include a:

- Growing tank,
- Sump of particulate removal device,
- Biofilter,
- Oxygen injection with U-tube aeration,
- Water circulation pump.

Depending on the water temperature and fish species selected, a water heating system may be necessary. Ozone and ultraviolet sterilization also may be advantageous to reduce organic and bacteria loads.

Water Supply

A good supply of water, adequate in both quantity and quality, is essential to a successful fish farming enterprise, RAS or otherwise. Ground water obtained from deep wells or springs is the best source of water for fish culture. It generally is free of pollutants and has relatively high hardness levels, which are beneficial under some circumstances. Municipal water supplies also can be used after chlorine, fluoride, and other chemicals are removed.

Other sources of water, particularly surface waters from streams, rivers, ponds, and lakes, are not recommended for fish culture. Surface waters may contain fish diseases, parasites, pesticides, and other pollutants that can kill or slow the growth of fish. Testing the quantity and quality of the available water supply is one of the first steps for a prospective fish farmer to take to insure an adequate supply of high quality water.

Because RAS recycle most of their water, they consume considerable less than other types of culture and are especially well adapted to areas with limited water supplies. The required quantity of water needed to grow fish varies with the species of fish selected, size of the culture system, and investment size. As a general rule, a minimum water volume of 1-5 gallons is needed for every pound of fish reared and minimum water flows of 10-25 gallons per minute (more for trout) are needed to grow 50,000 pounds of warm water fish per year.

Open vs. Closed Systems

Tank culture systems are referred to as recirculating (closed) systems because they recycle or reuse water. No system is ever completely "closed," because some water must be added periodically to replace evaporative loses and that used to flush out waste materials. Some water change is necessary since no filter is 100% effective. Nevertheless, RAS can operate efficiently by occasionally adding only a relatively small amount of water on a daily or weekly schedule.

Open (flow-through) systems refer to those that simply allow water a single pass through the system before it's discarded. Flow through systems can also be used in indoor tank culture of fish if an abundant and continuous supply of high quality water is available. Trout farmers typically use open systems rather than recirculating ones, because trout require large volumes of high quality, cold water. Open systems "leak" most of the water quickly, whereas closed systems "leak" water slowly. RAS are more suitable for warm water fish such as channel catfish, striped bass, and tilapia that can tolerate lower water quality conditions and higher temperatures than trout. They also are now being used to rear marine species such as redfish, oysters, clams, and soft-shell blue crabs.

Fish Culture Tanks

Fish can be grown in tanks of nearly every shape and size. Fish tanks typically are rectangular, circular, or oval in shape. Circular or oval tanks with central drains are somewhat easier to clean and circulate water through than rectangular ones. Rectangular tanks are usually built with or set upon inclined floors to facilitate cleaning and circulation.

Rearing tanks range in size from 500 to 500,000 gallons capacity. The size of the tank depends on a variety of factors including: stocking rates, species selected water supply, water quality, and economic considerations. The tank must be designed to correspond with the capacity of other components of the system, particularly size of the biofilter and sump so that all parts of the system are synchronized.

Tanks can be constructed of plastic, concrete, metal, wood, glass, rubber and plastic sheeting, or any other materials that will hold water, not corrode, and are not toxic to fish. Smooth surfaces on the inside of the tanks are recommended to prevent skin abrasions and infections to the fish, and to permit cleaning and sterilization.

Light weight, durable, plastic tanks can be conveniently moved and readily cleaned when necessary, but they require special support to prevent stretching when filled with water. Stainless steel also is a good tank material, but can be expensive. Marine-grade plywood tanks are inexpensive, but leak if not properly sealed and are not as durable as tanks of other materials. Concrete tanks may be the most economical to build, but they are relatively permanent and immovable structures once constructed. Non-toxic plastic or rubber liners can but used over frames made of wood, metal, concrete, or other materials.

Biofiltration

The biological filter (biofilter) is the heart of the RAS. As the name implies, it is a living filter composed of a media (corrugated plastic sheets or beads or sand grains) upon which a film of bacteria grows. The bacteria provide the waste treatment by removing pollutants. The two primary water pollutants that need to be removed are fish waste (toxic ammonia compounds) excreted into the water and uneaten fish feed particles. The biofilter is the site where beneficial bacteria remove (detoxify) fish excretory products, primarily ammonia.

Ammonia and Nitrate Toxicity

Ammonia and nitrite are toxic to fish. Ammonia in water occurs in two forms: ionized ammonium (NH_4^+) and unionized (free) ammonia (NH_3). The latter, NH_3, is highly toxic to fish in small concentrations and should be kept at levels below 0.05 mg/l. The total amount of NH_3 and NH_4 remain in proportion to one another for a given temperature and pH, and a decrease in one form will be compensated by conversion of the other. The amount of unionized ammonia in the water is directly proportional to the temperature and pH. As the temperature of pH increases, the amount of NH_3 relative to NH_4 also increases.

In addition to ammonia, nitrite (NO_2) poisoning of fish also is an imminent danger in RAS. Nitrite levels should be kept below 0.5 mg/l. Brown blood disease (methemoglobinemia) occurs in cultured salmon and channel catfish when hemoglobin is oxidized by nitrite to form methemoglobin (a respiratory pigment of the blood that cannot transport oxygen). The disease can occur at nitrite concentrations of 0.5 mg/l or greater. As the name implies, the blood has a characteristic chocolate brown color. Adding salt (NaCl) at a rate of 1 pound per 120 gallons of water (a chloride to nitrite ratio of 16:1) will suppress this disease in soft water; a ratio of 3:1 is effective in hard water.

Calculating Ammonia Loading

The amount of ammonia excreted into a tank depends on a number of variables including the species, sizes, and densities of fish stocked and environmental conditions (temperature, pH). Ammonia loading can be roughly estimated from the biomass (weight) of fish in the tank or it can be based on the weight of feed fed each day.

On the average about 25 mg (milligrams) of ammonia per day is produced for every 100 grams (3.5 ounces) of fish in the tank. Therefore, in a tank containing 1,000 striped bass fingerlings each weighing 75 g (75,000 g total fish weight), the daily ammonia load produced by all the fish would be 18,750 mg (18.8 g). To remedy excessively high ammonia levels, add freshwater, eliminate feeding or reduce the density of fish in the tank.

Ammonia loading also can be estimated based on the total amount of feed fed. For manufactured fish feed with standard protein levels of 30 to 40 percent, simply multiply the total weight of the feed (in grams) times 25. For example, if the fingerling stripers are fed 1 pound (454 grams) of pelleted feed per day, the amount of ammonia produced per tank would be about 11,350 mg per day.

Nitrification

Ammonia is a poisonous waste product excreted by fish. Since fish cannot tolerate this poison, detoxifying ammonia is fundamental to good water quality, healthy fish, and high production.

Detoxification of ammonia occurs on the bio filter through the process of nitrification. Nitrification refers to the bacterial conversion of ammonia nitrogen (NH_3) to less toxic NO_2, and finally to non-toxic NO_3. The process requires a suitable surface on which the bacteria can grow (bio filter media), pumping an continuous flow of tank water through the bio filter, and maintaining normal water temperatures and good water quality.

Two groups of aerobic (oxygen requiring), nitrifying bacteria are needed for this job. Nitrosomonas bacteria convert NH_3 to NO_2 (they oxidize toxic ammonia excreted by fish to less toxic nitrite), the Nitrobacter bacteria convert NO_2 to NO_3 (they oxidize toxic nitrite to largely nontoxic nitrate).

Nitrification is an aerobic process and requires oxygen. For every 1 milligram of ammonia converted about 5 milligrams of oxygen is consumed, and additional 5 milligrams of oxygen is required to satisfy the oxygen demand of the bacteria involved with this conversion. Therefore, tanks with large numbers of fish and heavy ammonia loads will require plenty of oxygen before and after the bio filtration process.

Nitrification is an acidifying process, but is most efficient when the pH is maintained between 7 and 8 and the water temperature is about 27-28 C. Acid water (less than pH 6.5) inhibits nitrification and should be avoided. Soft, acidic waters may require the addition of carbonates (calcium carbonate, sodium bicarbonate) to buffer the water. The addition of a salt as a therapeutic in striped bass as freshwater bacteria temporarily adjust to alteration in salinity.

Bio Filter Design and Materials

A bio filter, in its simplest form, is a wheel, barrel, or box that is filled with a media that provides a large surface area on which nitrifying bacteria can grow. The bio filter container can be constructed of a variety of materials, including plastic, wood, glass, metal, concrete, or any other nontoxic substance. In small-scale systems, some growers have used plastic garbage cans or septic tanks. The size of the bio filter directly determines the carrying capacity of fish in the system. Larger biofilters have a great ammonia assimilation capacity and can support greater fish production.

A bio filter must provide sufficient surface area for the colonization (attachment) of nitrifying bacteria. It needs to provide a large surface area to support bacterial populations at densities adequate to reduce the load of waste products (ammonia) excreted by the fish population in the tank. It is essential that the water flowing through the bio filter come into direct contact with the bacteria film growing on the surface media for a time period sufficient to allow the bacteria to convert toxic NH_3 and NO_2 to less toxic NO_3. Careful calculation of the flow-through rates (turnover or contact time) and size (volume and depth of the bio filter) is fundamental.

The bio filter media can be corrugated plastic, Styrofoam or glass beads, lava rock, sand, gravel, or similar material that supplies large surface area. The quality and quantity of surface area of the media provided for nitrifying bacteria are important determinants of the efficiency of the bio filter. The ideal bio filter media has high surface area for dense bacterial growth, sufficient pore spaces for water movement, clog resistance, easy cleaning and maintenance characteristics. We use and recommend plastic because it's lightweight, flexible, and easy to clean, but it can be expensive.

Bio Filter Sizing

The bio filter in any RAS design must be sized to correspond with the other system components. Important factors that must be considered in designing a bio filter are: media surface area (square feet of surface for bacteria attachment), ammonia leading (ounces of ammonia that need to be converted per day per square foot of media area), and hydraulic loading (gallons of water per day per square foot media surface).

Types of Bio Filters

Bio filters can be configured in many ways. The two general categories are submerged bed filters and emerged bed filters. Submerged bed filters can have fixed (immobile) media in which the water flow can be upward, downward or horizontally through the media.

The fluidized bed reactor (FBR) is a commonly used submerged bed filter. The FBR consists of fine particles (sand, dense plastic, glass beads, minerals, etc.) in a container through which upwelling water flows thereby "fluidizing" or suspending the media in the water column. FBRs offer large surface area per unit volume and theoretically greater nitrification. However, as with other submerged bed filters, all of the oxygen needed for conversion of ammonia to nitrate must be dissolved. Submerged bed filters often need supplemental aeration both before and after the water passes through the filter. If the inlet dissolved oxygen is low, the efficiency of ammonia conversion is reduced.

Emerged bed filters are of two basic types: Trickling filter (TF) sometimes called packed columns, and rotating biological contactors (RBC). These filters have the advantage of not requiring the addition of oxygen prior to water entering the filter. In fact, these filters frequently supply al the oxygen used to support fish respiration. For this reason these types of filters are often employed in RAS.

The trickling filter is designed to have water slowly cascade down through the media column, which is suspended above the water. Water enters the column (which is filled with biofilter media) from an overhead spray pipe and trickles down through the bio filter media where nitrification occurs. The waterfall action of this filter adds oxygen to (aerate) the water.

Trickling filters can become clogged and lose efficiency. When a filter accumulates too much organic material it can experience "ponding" and "short coming." In severe instances the filter can cease functioning altogether.

The rotating biological contractor (RBC) has a water-wheel configuration consisting of plastic media attached to a central axle which spins slowly, moving the media through the water in the RBC containment vessel. Advantages of the RBC are that it is self-aerating and self-cleaning. Once established, it tends to be very stable and can operate for years without failure.

The Virginia Tech system uses the RBC-type bio filter. The RBC has three circular stages (6' diam x 1' thick, 1,950 ft2 per stage, 5,850 ft2 total) mounted to a central shaft. The shaft is rotated at 2 rpm by a ¼ hp gear motor. The bio filter media and tank (525gallon) are divided into three partitions or stages so that filtered water passes in sequence form one to another. Staging the RBC allows greater nitrification efficiency for a given amount of media or RBC size.

Bio Filter Start-up

Efficient bio filtration depends on colonization of the bio filter media by nitrifying bacteria. Complete colonization of a bio filter is dependent on a number of environmental conditions and may take one to three months. Inoculation of a new tank with seed bacteria from a existing system can reduce start-up times and facilitate full efficiency. The addition of commercially available "specially selected strains" of bacteria has not probed to speed bio filter start-up as many advertisements claim. Cooler water temperatures, below 7 °F (21 °C), can reduce bacterial activity, slow the bacterial colonization, and diminish the effectiveness of the biofilter.

Recirculation Rates (Turnover Times)

The recirculation rate (turnover time) is the amount of water exchanged per unit of time. This can easily be determined by dividing the volume of water in the tank by the capacity of the pump (in gallons per minute). For example, the turnover rate in a 2,500 gallon tank system circulated with a water pump rated at 45 gallons per minute (63,360 gallons per day) would be 25.3 tank volumes per day (a rate of slightly more than one volume per hour).

Increasing the number of turnovers per day would provide increased bio filtration, greater nitrification (bacterial contact), and reduced ammonia levels. Most fish production recirculation systems are designed to provide at least one complete turnover per hour (24 cycles per day).

Compartmentalization

The ability of isolate the components of the system (bio filter, fish tank, and sump) is an important design feature, particularly critical when it becomes necessary to do filter maintenance or to treat the fish with chemicals and drugs. Cleaning and declogging static bio filters can pose difficult problems, particularly if there is no provision for shutting down the system for maintenance. Some therapeutic chemicals and drugs used to treat such fish may be harmful to nitrifying bacteria on the biofilter. A sudden drop in the efficiency of the bacteria can result in toxic NH_3 concentrations and fish kills.

Other Filters

Other types of filtration (mechanical and chemical) are available and can sometimes be used to supplement the efficiency of bio filters in removing ammonia in fish production systems. Most of these measures are useful only to temporarily control ammonia and nitrite in small systems.

In chemical filtration, water is pumped through a chemical media of activated carbon, zeolite, or other substances. These chemicals have microscopic pores that trap ammonia ions and remove them from the water. The familiar activated charcoal filter, popular in aquaria, can be incorporated as an auxiliary filter to support bio filtration in fish production systems, but this form of filtration requires periodic replacement with large quantities of relatively costly activated charcoal.

Zeolite filters are frequently used to remove NH_4 (and indirectly NH_3) at an estimated rate of 1 mg NH_4 per 1 g of zeolite. The use of zeolite requires regular and constant pumping of water through the filter and regular replacement of large quantities of expensive zeolite. Zeolite can be recharged with a salt solution (10%) and reused, but salt water disposal then becomes an environmental problem, particularly in inland waters.

Sump

A sump (clarifier tank) is used to prevent the excessive accumulation of fish excretory products and waste feed. Waste products increase the biological oxygen demand (BOD), decrease the dissolved oxygen content, lower the carrying capacity (density of fish) that can be reared, and may result in off-flavor in fish products. Accumulation and decomposition of waste material results in the production of toxic compounds such as ammonia (NH_4, NH_3, NO_2) and hydrogen sulfide (h_2S) that can be hazardous to fish health.

The clarifier tank is designed as a settling basin (large volume tank with a slow flow rate to increase sedimentation). Its purpose is to concentrate and remove suspended solids (fish feces, uneaten feed particles) before they clog the bio filter or consume valuable oxygen supplies. The clarifier should be a separate tank, isolated from the fish tank and the bio filter, so that it can be cleaned periodically (daily) as needed. To increase the efficiency of the clarifier, various filters (plastic filters, sand filters, metal screens) can be inserted into the sump tank.

In the RAS, water flow in the sump tank is directed upward through a multi-tube clarifier (corrugated plastic filter) which removes large, settle able particles from the water column by gravity (particles with densities great tan water will sink). The waste material sinks, collects, and is concentrated in the bottom of the sump where it can be drained out. The sump tank should have a v-shaped bottom to concentrate waste particles and facilitate cleaning. Another technique for the removal of waste particles for suspension in the water column includes the use of a hydro clone, a cone shape tank that uses centrifugal force to increase gravitational force and enhance waste particle settling.

The size of the clarifier tank needs to be fully integrated with the sizes of the fish tank and biofilter and also with the turnover rate of the system (pump size). Volume of the sump and flow rates through the sump must be adjusted to maximize sedimentation of suspended particles.

Oxygen Management

Successful fish production depends on good oxygen management. The addition of oxygen in a pure form or as atmospheric air (aeration) is essential to the survival (respiration) of fish held in high densities, the survival of aerobic, nitrifying bacteria on the bio filter and, for the decomposition (oxidation) of organic waste products. Supplying sufficient oxygen to sustain healthy fish and bacterial populations and to meet the biochemical oxygen demand (BOD) for fish waste and unconsumed food is critical. Maintain oxygen levels, near saturation or even at slightly super-saturation at all times. Low oxygen levels will reduce growth, feed conversion rates, and overall fish production.

The amount of oxygen needed in RAS depends on a number of factors. Oxygen demand is directly correlated with the density of fish in the tanks, feeding rates, water temperatures, flow rates, and nitrification. It is also a function of physical conditions such as water temperature and water volumes. Increasing dissolved oxygen concentrations through oxygen injection, aeration, and increasing water flow rates (turnover times) are ways to increase the density (carrying capacity) of fish that can be held in tanks of fixed size.

Atmospheric oxygen can be added to the tanks by surface agitation with aerators or by large blowers. Surface aerators may not be cost effective or efficient in evenly distributing oxygen throughout large commercial-scale systems. Blowers can be effectively used to supply oxygen and also to mechanically rotate RBS.

Pure Oxygen

Pure oxygen injection systems are increasingly being used in aquaculture. They are particularly useful in maintaining oxygen-saturated conditions in recirculating systems with high densities of fish. Pure oxygen can be delivered and stored in a tank as liquid oxygen or it can be produced on-site by an oxygen generator. Bottled oxygen gas also is sometimes kept as an emergency backup system for RAS, but this alternative usually is too expensive and bulky to be practical.

Liquid oxygen technology is relatively simple, efficient, and cost-effective; especially if purchased in bulk quantities and if the site is located near a reliable supplier. A liquid oxygen system consists of a storage tank for the liquid gas, vaporizers to turn liquid oxygen to gas, and supply lines to the fish tanks. It conveniently requires no external power supply and is therefore free of power failures and the consequent fish kills. Most growers rent or purchase a liquid oxygen storage tank of a size sufficient to provide two to four week supply of oxygen. The size of the tank corresponds with the fish production capacity of the system.

Oxygen generators (pressure swing adsorption systems, PSAs) are particularly advantageous at remote sites where liquid oxygen deliveries would be costly and expensive. Generators produce oxygen by using electric energy. They are expensive to purchase and operate, and they are subject to power failures. Selecting the proper size of the oxygen generator and carefully calculating the cost of electric power needed are important considerations.

Oxygen Diffusion

Effective diffusion of pure oxygen gas into a liquid (water) can best be accomplished using a U-tube oxygenation, counter-current flow injectors, or micro-bubble devices (tubes or fine wetstones). The purpose is to dissolve much of the oxygen injected so that it is available to the fish, rather than wasted by bubbling out of solution to the atmosphere.

A shallow well (40 feet deep) into which tow concentric pipes (one outer 6 inch and an inner 3 inch diameter pipe) are suspended, the water, injected with oxygen at the top of the U-tube, flows down the inner pipe, up the outer pipe, and back into the fish culture tank saturated with oxygen. The U-tube increases oxygen transfer because of the longer contact time and greater atmospheric pressure (two atmospheres at the bottom of the U-tube) of the entrained oxygen gas bubbles and water. Although the U-tube aerator is not essential, oxygen enrichment (entrainment) is significantly increased by this cost-effective method.

Ozone Sterilization

Ozone (O_3) is a naturally occurring gas (upper atmosphere) that consists of three atoms of oxygen. It is a powerful oxidizing agent that can be used to break down compounds. Ozone must be used with caution since it is directly toxic to aquatic life and may form harmful byproducts (hypochlorite, hypobromite). Careful redox potential measurements and special injection equipment apparatus are needed to determine and control ozone applications.

Carbon Dioxide

In addition to toxic ammonia, carbon dioxide tends to concentrate in intensive fish production systems. As carbon dioxide increases, the pH of the water decreases, and fish respiration is affected. Carbon dioxide levels should be maintained at levels less than 30 mg/l for good fish growth. Some carbon dioxide is beneficial since it reduces pH and mitigates ammonia toxicity. Carbon dioxide removal can be accomplished with any device (RBC, packed column) that increases air-water contact.

Emergency Systems

Fish farmers using RAS usually have made a substantial economic commitment in the system, feed, and fingerling fish. The standing crops of fish in the tanks at any particular time represent costly inventory that become increasingly valuable as they approach marketable size.

Fish kills caused by suffocation resulting from power failures (even short-term ones) are not uncommon. The oxygen demand in tanks with high densities of large fish is high. During emergency power failures, the grower generally has a response time of 10 minutes or less to avoid a total kill. Sub lethal oxygen concentrations can be equally harmful because they cause slow growth, increase the susceptibility of fish to disease and parasites, and decrease production and profitability.

Fish farming in RAS is risky business where "an ounce of prevention" is a worthy and important consideration. Fish kills from oxygen deficits as a result of power failures constantly threaten growers with financial ruin. Protection by designing alarm systems, or using backups and alternatives for emergencies is essential.

An alternative bottled oxygen gas or liquid oxygen system is suggested for growers who rely primarily on oxygen generators. The RAS uses liquid oxygen, but also has a secondary, standby surface agitator system in each tank and a backup electric generator for emergencies.

An emergency water supply reservoir consisting of a large volume of high quality water (oxygenated, de-chlorinated) and the proper water temperature stored in a standby tank provides a good margin for safety. Sudden declines in water quality or outbreaks of disease can be effectively controlled in certain instances by rapidly flushing growing tanks with freshwater, by quickly altering water temperatures.

Production Considerations

Fish Species Selection

RAS can be used to rear nearly any species of fish, freshwater or marine or other aquatic animal. Hybrid striped bass, channel catfish, and tilapia are the most common freshwater species grown in these intensive culture systems. Red drum (redfish) and soft shell blue crabs are the main saltwater species being reared RAS. Because RAS requires high capital investment, it is best designed to grow high-value species such as striped bass for food or stocking in private recreational angling waters or aquarium fish.

Certain fish species are easier to grow and more available than others. Two species dominate the U.S. aquaculture industry today: channel catfish and rainbow trout. Their popularity is a result of the desirable traits that they express. For example, their culture requirements are well known, the demand high, and the economics viable. Small (fingerling) fish are available for grow out nearly year-round, they tolerate crowding and stress, and disease resistant, and grow rapidly to a marketable size.

Recirculating systems generally are not suitable for rearing Coldwater species such as salmon and trout. Trout and salmon require clean, cold water and generally do not thrive in warm, recycled waters. However, a wide variety of fish, particularly catfish and hybrid striped bass, and other aquatic species, including shrimp and mussels are good candidates for farming in recirculation systems.

Feeds and Feeding

A complete feed, containing all the essential minerals and vitamins for healthy fish growth, and formulated specifically for the fish species is being reared, is necessary for fish production in RAS. Do not substitute other animal feed for fish feed. Even different fish species have different nutritional requirements, particularly the quality and quantity of protein needed, that must be met to optimize growth.

We recommend feeding a commercial feed of dry, floating pellets so that the feeding activity and health of the fish can be easily observed at the water surface. The size of the pellet should correspond with the size of the fish. Feed the largest pellet that the fish will readily swallow in order to maximize consumption and minimize waste. To keep feed fresh, order only a limited supply and store it in a cool, dry area, free of insects and rodents. In case of a liability problem with contaminated feed, it's a good idea to freeze small samples of each new batch of feed purchased for subsequent analysis if necessary.

Cultured fish generally are fed 3 to 5 percent of their body weight or all the feed that they can consume in a short period of time, say five minutes. Feed remaining in the tank after five minutes is seldom eaten and overfeeding can seriously degrade water quality. A good, quick indicator of problems with water quality or disease is when fish go off feed or refuse to eat. If fish suddenly stop feeding, immediately check for high ammonia levels, low oxygen concentrations, diseases, or other problems. Reduced feeding rates occur at very high and low water temperatures.

To maximize growth, feed on a regular schedule at the same time each day. More frequent feedings several times per day) have resulted in better growth rates and feed conversion efficiencies than a single daily feeding. Distribute the feed as uniformly as possible throughout the tank to prevent uneven growth and stunting. At the RAS Aquaculture Facility, the fish are hand fed. Automatic feeders and demand feeders can be used in commercial operation to reduce labor costs.

Warm Temperature

Water temperature strongly influences feeding and growth rates of cultured fish. The water temperature preferences of most cultured fish are well known and can be maintained year-round in RAS. In general, cultured fish species can be classified as either cold water species that prefer temperatures of 50-65 °F, (trout), cool water species prefer temperatures of 65-80 °F, (yellow perch), and warm water species (channel catfish) that prefer temperatures 80-90 °F. Water temperature also influences the water quality processes occurring in recirculation systems. For example, the optimum water temperature for bacterial nitrification activity is 85 °F, which may or may not be optimal for the fish species being cultured.

Energy conservation is one of the major advantages of recirculation aquaculture systems. Once the tank water is heated to the optimal temperature for fish growth, only a small amount of heat energy is required to maintain the temperature. However, heat loss from the building and from the water via evaporation, splash-out, or waste water must be minimized to assure economic success.

A unique thermal property of water is its high specific heat – it heats and cools slowly. Therefore, once the optimal water temperature for fish growth is reached, only a small amount of energy is necessary to maintain the best thermal condition. Heat losses through convection and conduction are minimal in a well-insulated building.

Heat can be provided by heating the air in the building, or directly heating the water, and by using heat exchangers. Water temperature in the RAS is supplied by is supplied by heating the air in the building with four propane space heaters. The room temperature is generally kept 2-4 degrees above that of the water to reduce condensation in the building. The RAS is housed in a specially constructed 110- x 50- x 16-ft high farm utility building. The interior and exterior walls and ceiling are covered with 19 gage corrugated aluminum siding and insulated with R-19 and R-30 fiberglass blanketing. At times, room temperatures and working conditions can become hot and humid, especially when growing warm water fish in the summer. High temperatures and humidity can be controlled through ventilation with an electric fan.

Directly heating the water is another alternative, but this strategy requires expensive insulated tanks with lids that maintain the heat, but inhibit feeding and observation of the fish. Alternatives such as solar heating and heat exchangers are being considered.

Rice-Fish Culture

A rice-fish system is an integrated rice field or rice field/pond complex, where fish are grown concurrently or alternately with rice. Fish may be deliberately stocked (fish culture), or may enter fields naturally from surrounding water ways when flooding occurs (rice field fisheries), or a bit of both. Fish yields can range widely from of 1.5 to 174 kg/ha/season depending on the type of rice fish system, the species present, and the management employed.

Rice-fish systems allow the production of fish and other aquatic animals, from the same rice field area and generally without causing reductions in rice yields. This source of animal protein may be important for household nutrition and farm income.

Fish Species

The most common indigenous fishes [common name (Genus)] found in Asian rice fields are the following:

- White fish (small plant or plankton eating species) such as Danios (Rasbora), Barbs (Puntius), Snakeskin Gourami (Trichogaster), and Half beaks (Xenentodon).
- Black fish (often carnivorous air breathers that can survive low or no oxygen levels) such as Snakehead (Channa), Catfish (Clarias), Climbing Perch (Anabas), Spiny eels (Mastacembelus), and Sheatfish (Ompok).
- Introduced exotic fish species such as Common Carp (Cyprinus), Tilapia (Oreochromis), and Silver carp (Hypophthalmichthys).
- Other wild aquatic species such as crabs, shrimp, snails, and insects may also be harvested.

Site Selection

Certain criteria should be followed while selecting the site (land) for integrated rice-aquaculture. The selected area (place) should have an optimum rainfall of 80 cm in a year; the lands with uniform contour and high water retaining capacity of land are considered for site selection. The site selection for rice cum aquaculture farming is low lying area where water flows easily and is available at any time in needs. The fertile soil rich in organic manure and with high water holding capacity is used for rice fields. In general, soils with medium texture and loam with silty clay are most preferable for RAF.

Rice Field Selection

The topography and contours of the land will make the difference in preparation of rice field. There are three types of rice fields identified such as:

Perimeter-type field: The field in this type prepared with a moderate elevation and ground sloping on all sides into perimeter trenches which facilitates easy drainage. The middle of the land is the growing area for rice.

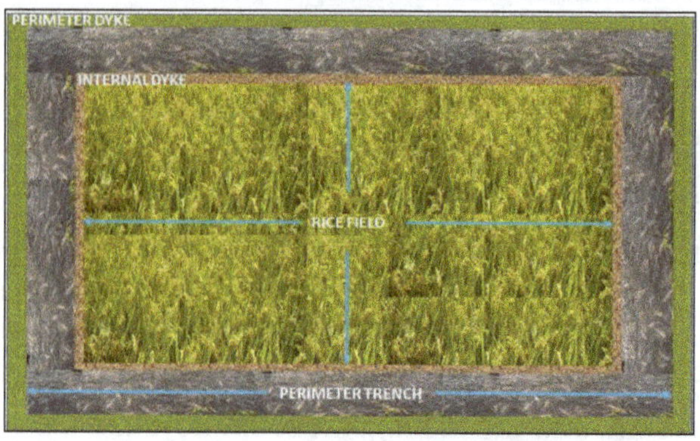

Figure: Perimeter-type rice field used for rice-aquaculture integrated farming.

Central pond-type field: The pond is prepared in the middle of the rice field. This is the easiest way to produce trench for integrated aquaculture and rice production.

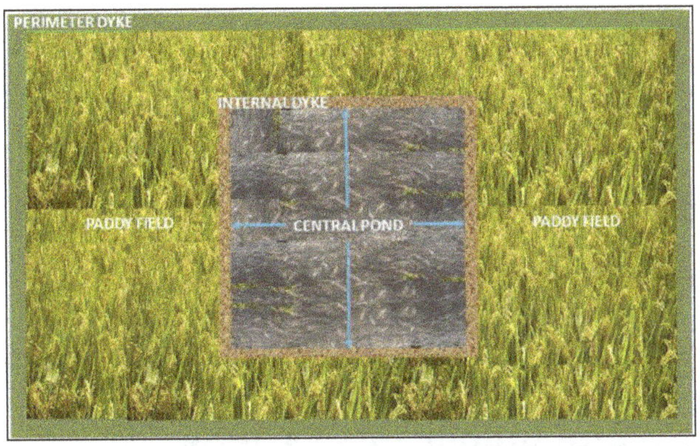

Figure: Rice field with central pond system for rice-aquaculture farming.

Lateral trench-type field: Trenches were prepared on one or both sides of rice slopes.

Figure: Lateral trench-type rice-aquaculture integrated farming.

Besides these three types, there are many types of trenches prepared for integrated rice-aquaculture. Different other types of trenches in the rice field are presented in Figure below. The ideal ratio for making an integrated RAF is explained well, for example if a 1 hectare of land of integrated system is about 125 m × 125 m. In this, 0.67 hectares is the area to be used for rice, i.e. 102.5 m × 102.5 m, whereas 7.5 m × 440 m for four sides of the field which is equal to 0.21 hectares were used for aquaculture (fish, crustacean and other aquatic animal culture). About 0.12 hectares of area, i.e. 3.75 m × 485 m, will be used as embankment area. The other area of about 0.04 hectares is used for planting fruits.

Figure: Different types of trenches in usage for rice-aquaculture integrated farming. (A) Diagonal trench, (B) crossed trench, (C) Y-shaped trench, (D) peripheral and one central longitudinal trench, (E) two equidistant transverses with peripheral trench and (F) latticed trenches.

Rice and Aquaculture

RAF had been in practice since years due to the constant profits of integrated culture. However, the generally practiced rice-aquaculture is categorized into three major types, namely:

- Synchronous/combined/simultaneous farming.
- Sequential/rice-aquaculture rotation/alternate farming.
- Relay farming.

Synchronous/Combined/Simultaneous Farming

In this type, the rice and aquatic culture species are grown simultaneously in the same field. Both the rice and cultured species are harvested at the same time. The synchronous rice-aquatic farming method has its own importance since it produces aquaculture protein with no additional cost; the cultured aquatic species secretes ammonia for rice growth besides destroying the weeds, tillering, enrichment of minerals (digging activity) and utilization of unused aquatic feed by the crop plant. Due to this the rice yield increases 5–15% than the conventional methods.

This method of integrated culture is an old one and is familiar in the Asian countries. Usually in this type, 5 days after transplantation or after proper rooting of rice seedlings, the aquatic culture species seed is released into the trenches/pond (also called as stocking). The stocking time after transplantation of rice seedlings varies from aquatic cultural species to species. For example, stocking in the case of fry is after 10 days of transplantation, whereas for fingerlings after 3 weeks. The stocking procedure differs with the type of aquatic species growing (fingerlings or fish or prawn or crab or mollusk or any other aquatic species). In simultaneous farming, the rice selected is usually local variety harvested not less than 5 months by using organic manures like stocked cow/buffalo dong or green manure fertilizers without synthetic fertilizers. Moreover, the number of crops produced every year is two in this case. However, the combined rice-aquaculture farming has its own disadvantages such as high activity by predators of aquatic species (piscivore birds) due to less depth in trenches/ponds and restricted usage of herbicides and pesticides which causes harmful effects to growing aquatic species lowers rice yield; there may be 20–60% loss in aquatic species yield due to abrupt change in dissolved oxygen and temperature and limited space availability. Since having drawbacks in simultaneous rice-aquaculture farming, it was modified and established as alternative farming.

Sequential/Rice-Aquaculture Rotation/Alternate Farming

Rice-aquaculture rotation is a simple method of farming where aquatic species were grown after harvesting of rice in the flooded fields without removing the rice stubbles. The water levels were shallow in the case of rice, whereas for aquatic species, water depth is raised. In this method more than one ASp was raised. The depth for growing aquatic species in the field is raised well before the transplantation of rice seedlings with a moderate height of external dyke. However, the pesticides used for rice production degrades during interval between harvesting of rice and stocking of aquatic species. The rice stubbles in the water at the bottom of the field facilitate the growth of decomposing microbes which served as food for aquatic cultural species. The decomposition of stubbles enriches the water and soil with natural fertilizer utilized by the next cycle of rice where it helps to produce high yield.

The only limitation in alternate farming is about 20–60% loss of ASp due to piscivorous birds like herons and cormorants. Besides this, sequential rice-aquaculture farming has many advantages such as:

- No limitation for depth of water neither in rice cultivation nor in aquatic species culture.
- Maintenance of adequate water levels provides sufficient dissolved oxygen and water temperature.
- Rice stubble decomposition facilitates microbial food for aquatic species and fertilizer for the next cycle of rice for improved production.
- Useful for mono- and polyculture of aquatic species.
- Reduced attack of insect pest on rice fields due to interruption in their life cycle in rotation farming.

Relay Farming

The name 'relay' itself indicates it is a prolonged farming. The relay farming is a complicated system compared to synchronous and alternative rice-aquaculture farming and is a combination of these two methods. In this farming rice and aquatic species were started like synchronous farming, but the aquatic species are not harvested while rice species are harvested. It means that relay farming requires longer period for aquaculture. During rice harvesting the growing aquatic species were transferred to special ditches/ponds which are connected to channels/pools of the rice field and then restocking them in the rice field after filling up of water for further growth. It means this method requires additional rearing facility for transferring fish at the time of rice harvesting. Relay farming provides high amount of aquatic protein with high yield of rice crop with short duration. The drawback in this method is more investment and labor. The carp and prawn species are the most commonly used to produce protein in relay farming.

Suitable Aquatic Species

Worldwide integrated farming of aqua with irrigation is becoming popular to produce more yield and makes the farmers economically strong. It is well practiced in the rain fed areas mainly in countries producing rice. The 86% of rain fed areas in Thailand produce rice as is the case with Lao PDR and Cambodia. Various types of aquaculture in the rice fields are in practice. The major aquatic species that come under finfish and shellfish are the cultural species used for integration with rice fields. Among finfish and shellfish, the two types of cultures brackish and freshwater species have importance to grow in RAF. But the selection of this species purely depends on the variety of rice growing. It is very particular that only certain rice species are productive in brackish water. The rice varieties of holding salinity tolerance and floating-stem (long-stemmed) were preferred for brackish water RAF. The brackish water effect is tidal on rice species holding salt tolerance and long-stem, since these species have the capacity to withstand even in salt water. Besides rice variety the selection of fish or crustacean cultured also play a role in RAF. The freshwater prawns like Macrobrachium species and brackish water prawns like Penaeus monodon and other species grow well in the rice fields. Most species of fin fish can grow well in rice fields. However, the farmers have to be careful about the selection of aquatic culture species and the variety of rice for integrated culturing.

Studies were focused on the stability of rice field ecosystem with rice monoculture and rice-aquaculture. A survey by Xie et al., with farmers cultivating rice with or without integrated aqua farming, revealed no difference in the yield of rice between the two farming for a period of 6 years. However, the temporal variation in the yield of rice is influenced by many factors such as year-round and year-to-year changes in climate, pest incidences, rice variety cultivation and the amount of pesticide used until harvesting. The usage of large amount of pesticides is responsible for temporal stability of rice yield in the case of rice monocultures, whereas rice yield stability is partly on the presence of aquatic species in rice-aquaculture farming, since fish-aquatic cultural species act as bio control agent. In the modern agriculture, the stability of rice yield is maintained through the extensive usage of pesticides and fertilizers. In contrast, the stable yield of rice is highly maintained by exploiting synergies between species to minimize usage of chemicals and is suggesting to the modern agriculture that rice monocultures need to be improved with addition of species for facilitating positive interactions between species and components. The interactions between various components of RAF are presented in figure.

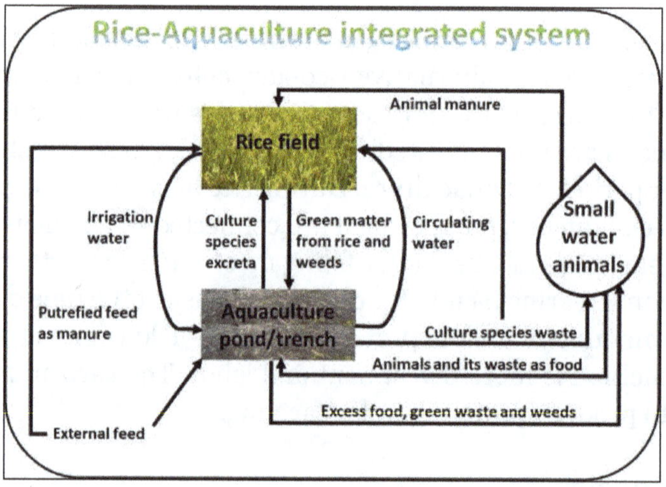

Figure: Interaction of different components in integrated rice-aquaculture systems.

The type of method for integrated culture mostly depends on local conditions of the culturing area. The climatic and other local conditions, cultured aquatic species and rice varieties, method of culturing, fertilizers used and supplementary feed are the principle factors that influence the quality and quantity of yield. The culture method selection depends on the source of aquatic species (ASp) and its interaction with the growing rice which was well described by Coche seven decades ago.

The integrated rice-aquaculture is a mixed farming, where various rice varieties are grown along with selected aquatic cultural species depending on the local biological and physical factors. However, the selection of one aquatic species or more influences the yield of aquaculture protein. Because of this, the selection of aquatic species plays the most important role for the profit to the farmer and helps to improve the economy of the country. The simple economic benefit of shrimp/prawn, crab and fish with rice cultivation is well defined. Besides profit rice-aquaculture generates the highest employability in rice-prawn culture, followed by rice-crab, rice-shrimp, rice-carp and rice-tilapia.

Suitable Crustacean Types

There are a number of crustacean species that grow in the rice fields as an integrated culture. Besides crabs, there are other crustacean species like prawns, shrimps, lobsters and crayfishes that can also be grown in the integrated system along with rice with proper management, which is in practice in the coastal areas. Since the most profitable crustacean protein in today's world market is from shrimps or prawns, the production of these species is more profitable to the farmers especially the Indian and Asian conditions.

Shrimp/Prawn Production

Shrimps are grown in brackish water, whereas prawns grow in fresh water. The shrimp production is done with the fields of holding salty water located in west coast and deltaic areas of eastern India where the salt-resistant rice varieties are grown. Due to limited availability of salt water, only one crop (integrated rice-shrimp farming) is grown mostly in the months of July to September every year. The channels are arranged to control the water supply into the field or pond during harvesting of rice and the shrimp is protected in the ponds/trenches. At this stage, rice field consists of substantial number of shrimp larvae having less water. The shrimp harvesting in this case is usually done after 3 months of rice harvest. To clearly say the method followed to culture rice and shrimp is through alternative system (semi-intensive monoculture) with few modifications and balanced diet supplied during the culture. The average production rate of shrimp per hectare in this method is 1440 kg. Penaeus monodon is the most common species grown in this type of culturing. Penaeus merguiensis, Penaeus indicus and Metapenaeus ensis are alternative species to grow in this type of system.

The most popular species of freshwater prawn grown in the integrated rice-prawn culture is Macrobrachium rosenbergii (M. rosenbergii). The monoculture of M. rosenbergii in the rice fields is in practice in North Kuttanad, where almost 248 hectares are cultivated integrated rice-prawn culture in a year. Like rice-shrimp farming, the alternative system of culturing is adapted for rice-prawn culture (semi-intensive monoculture). The productivity of prawn monoculture with rice is around 700 kg/hectare. On the other hand, the polyculture of prawn (M. rosenbergii) along with fish (Catla catla, Labeo rohita and Ctenopharyngodon idella) in the rice fields is also in practice, where the productivity of prawn is 285 kg/hectare, which has almost two times less production of monoculture of prawn in the integrated system. The yield of *Penaeus monodon* with mixed cultures in the rice fields is about 2135 kg/hectare.

Crab Production

More than two decades ago, the role of indigenous technology in the production of rice-crab culture is well described. The crabs *Oziotelphusa senex senex* or *Parathelphusa hydrodromus* are commonly called as 'freshwater rice field crabs' since their occurrence is naturally in the rice fields. Besides this Eriocheir sinensis (Chinese mitten crab) and Scylla serrata (mud crab) are the major crab species used for rice-crab integrated farming. Mitten crab is a famous cultural species of China, whereas the mud crab is the most economic crab identified and grows throughout the world. Both these species grow in saline water and are cultured along with salt-resistant rice species in the case of integrated culture. Scylla serrata is popular in Asian continent. However, the crab culture is influenced by many factors, including environmental, biological, physical as well as socio-economic factors.

Crabs depend on naturally available food in the rice fields. Anyhow, supplementation of external feed is needed during fast-growing periods. The management of rice-crab culture is much easier than other RAFs. But the only management aspect that needs to be addressed is regulation of pesticide usage. Since the usage of pesticides is unavoidable for rice yield, it can be managed by selecting less toxic pesticides with a method of spray on rice stems and specific time of spray at which crabs stay in burrows. Proper water management prevents the crab from escaping from rice fields. The average crab protein produced in rice-crab system is approximately 600 kg/hectare. The systematic management of crab culture changes the farmers socioeconomic status than the shrimp/prawn culture alone in addition, these crabs are more eco-friendly and grow easily in polyculture with other species. The only limitation in crab culture is limited availability of wild seed and is overcome by establishing crab hatchery industry to protect natural broodstock for continuous supply of seed. The sequential or synchronous farming is the adaptable system for rice-crab culture, which depends on type of crab species growing.

Crayfish Production

Similar to crabs, crayfishes are the strongest crustaceans that tolerate unfavorable conditions than other species. The most common species used in rice-crayfish integrated system are Procambarus clarkii and Cherax quadricarinatus. The rice-crayfish culture is common in Southern United States (Procambarus clarkii) and in Australia (Cherax quadricarinatus). The adult crayfish is used as broodstock to get seed in the ponds. The broodstock is released into field after 1 week of rice seedling implantation. On the other hand, the crayfish seed is also obtained from hatcheries. Besides this the seed also obtained from the rice-crayfish integrated system after harvesting, where the adult crayfish produce juveniles which are used as seed for immediate culture. Unlike other crustaceans rice-crayfish culture is also done with sequential farming. In this system before 2 weeks of harvesting the rice, the water is drained to facilitate crayfish to make burrows. Watering the field after harvesting leads to growth of ratoon crop (occurrence of regrowth from rice stubbles) and facilitates the growth of insects, zooplankton, worms and molluscs, which promotes the direct growth of flushed out crayfish from the burrows. The optimum weight for harvesting crayfish is about 15–60 g. The average yield of crayfish in this system is about 1960 kg/hectare. Many other aquatic species can be integrated with rice culture. The typical integration of aquaculture with rice system followed in Asian countries.

Rice-Aquaculture Farming with Livestock

The profit of rice-aquaculture integrated culture created a scope for betterment of social-economic status of farmers. Using this as a concept, the farmers tried to integrate other livestock with RAF. There are many systems evolved in this way and mentioned in table. However, the livestock rearing with RAF is in primitive stage and has its own limitations.

Table: The livestock integrated systems with rice and/or aquaculture in Asian countries.

S. no.	Livestock integrated with rice-aquaculture system	Product produced from livestock
1.	Duck	Eggs and meat
2.	Golden apple snails	Meat
3.	Cows and buffalos	Milk and meat
4.	Pigs	Meat
5.	Sheep and goat	Milk and meat
6.	Poultry	Egg and chicken

Management of Rice-Aquaculture Farming

Individually the rice and aquaculture systems are separate entities but are mostly holding common requirements. Because of commonality in system management, the integrated RAF is prospective and profitable. However this integrated system has conflict in requirements and their management from one system to other. The modified interventions are required to overcome these problems at each and every aspect and step of system management.

Water Management

Water plays an important role in any agricultural system. Continuous water supply is badly needed for RAF. The source of water (river, pond, well or bore) is the first parameter one should think of before establishing the integrated system. Uninterrupted water supply required for rice up to its maximum tolerated levels (15–20 cm) is recommended and is good for maximum yield of rice. For the aquatic species, the depth of water in the ponds/trenches should be 65–70 cm and is sufficient to manage cooler and hotter areas for the better growth of aquatic species. Most of the times, crustaceans grow at the bottom whereas fish at the middle and top layers of the water.

Fertilizers and Chemicals

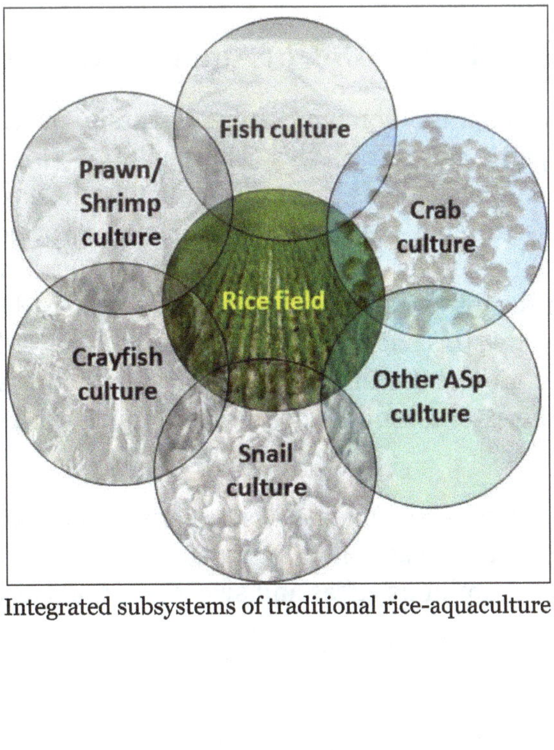

Figure: Integrated subsystems of traditional rice-aquaculture in Asia.

Organic and inorganic chemicals and fertilizers are essentially required for the growth of both rice and aquatic species. Besides nourishment of rice, external supplementation of reasonable amount of nutrients raised the levels of phyto- and zooplankton, which serve as food for aquatic cultural species. Aquatic species in the rice field enrich the soil fertility, which ultimately reduces the external supplementation of fertilizer and reduces the production cost. This proves that RAF requires less usage of fertilizers than rice alone. Alternatively yield of rice increases by the supplementation of nitrogen fertilizers at the time of land preparation instead of applying at the time of farming. Addition of ammonia during farming causes damage to ASp as it acts as toxicant in the water, whereas no change has been seen with phosphorus fertilizer application in soil and water.

Organic fertilizers benefit for the growth of both rice and ASp. Moreover, the additional supplementation of animal manure helps for the better growth of aquatic species at the time of land preparation, and in this way, any toxic effects of these manure to the aquatic species are avoided. Nevertheless, integrated rice-aquatic culture farming is good for poor and unfertilized soils where the aquatic culture species play the greatest role in making them fertile.

Besides fertilizers and nutrients, the control of pest and fungal diseases of rice is another task for production of higher yield of rice which usage is not at all good for the growth of ASp. In addition to reduced yield, the uncontrolled usage of pesticides and fertilizers declines the biodiversity. Prolonged misuse of pesticides and fertilizers over the years has also halted the development of inland fisheries and aquaculture.

In an attempt to reduce pesticide use, important changes have taken place in strategic approaches to plant protection. Integrated pest management (IPM) methods have brought ecological principles and social scientific perspectives into traditional crop management. These ecology-based pest control methods have resulted in markedly improved rice farming systems, which are not only higher yielding but also more sustainable. In addition to this, increased adoption of rice-fish farming, with fish as a natural control agent of pest organisms, provides a promising alternative for further developing ecological sound management strategies of the rice field environment.

Nutrition and Supplemental Feeding of Aquatic Cultural Species

The rice ecosystem is rich with natural flora and fauna. Naturally the rice ecosystem itself is a rich nutrient supplement for aquatic cultural species. The phytoplanktons and bacteria are the first developers in the rice field and later weeds (macro phyton, benthos and detritus), and zooplankton will develop. All these are directly consumed by the aquatic species. Usually, farmers enrich the natural nutrients of aquatic species by supplementation of fertilizers. An alternative to have natural supplementation in RAF is by raising a fern azolla. Azolla is the best natural feed for fish, but not for crustaceans. Anyhow, this naturally occurring food is not sufficient for growing aquatic culture species. The consumption and utilization of natural field nutrients vary from aquatic culture species to species. So, the external supplementation of nutrients is a must for attaining higher yield of aquatic protein.

The composition of external feed supplemented for crustaceans/other aquatic species is a mix of 40% of animal waste (fish, snail, clam and viscera tRAFh), 25% of plant waste (rice or wheat bran, vegetables, pumpkin, leguminous cakes and sweet potatoes) and 35% of terrestrial gRAFs and/or duck weeds. These materials are made as powder and used for making food pellets. But various types of external feed are in practice in many countries.

Temperature and Dissolved Oxygen

The fluctuations in the temperature and dissolved oxygen do not have much influence on the yield of rice, but it matters for the production of aquatic protein. In summer, the water temperature for ASp is regulated by rice plant shading during summer. Anyhow the low/moderate temperatures are best suited for good yield of ASp. To maintain adequate temperatures for the better growth of fish, prawn, carb, crayfish, snails and other ASp, high amount of water needs to be maintained continuously in the ponds/trenches/refuges. The only way to maintain temperature in the rice-ASp system is by controlled water system.

The dissolved oxygen is an important factor that affects the survival and growth of ASp, and in extreme conditions, it may lead to death. Unsystematic management of ponds/trenches leads to pollution, resulting in reduced dissolved oxygen. Basically the reduced dissolved oxygen changes the physiochemical properties of water; thereby pond/trench ultimately shows effect on growing fish. Most of the times, dissolved oxygen is reduced due to overloading of nutrients in the ponds/trenches and also causes increase in cyanobacterial bloom, which depletes dissolved oxygen and shows poisonous effects on ASp. The increase in photosynthesis in rice due to increased aeration is recorded with the fish movement in the rice field.

Predators and Other Factors

The birds are the predators for the ASp in rice-ASp system. Small birds may not cause damage to the system, but the larger ones reduce the ASp production to 20–60%. Aquavorous birds should be controlled by placing certain bird terrorizing signs or holdings in and around the rice-ASp system or making sounds to create panic and foreboding in them.

Advantages of Integrated Rice and Fish Farming

The following are the advantages of integrated rice and fish farming:

- Farmers can get extra income from rice-fish culture along with main crop yield.
- Rice-fish culture controls insects which are harmful to paddy (rice).
- Rice-fish culture reduces the risk of crop failure.
- Rice-fish culture is useful in controlling weeds.
- Rice-fish culture increase the production of rice as fish stir up soil nutrients which are very helpful for rice crop.

Disadvantages of Integrated Rice and Fish Farming

The disadvantages of integrated rice-fish culture are:

- Rice and fish farming requires restricted use of pesticides.

- This integrated system of fish and rice requires more water than paddy (rice) cultivation alone.
- Rice yield is less in this system as trenches are required about 45 cm below the paddy bottom. Due to these trenches, the paddy cultivation area will be reduced and it results in lower yields. Digging trenches makes difficult in draining the water.
- In rice and fish farming integrated system, water regulation is important and paddy fields cannot be dried while fish is stoked in the field.
- The fish size is small and total production of fish produced in paddy-fish culture is less when compared to open pond fish farming.
- Commercially rice-fish culture many not be feasible as fish must be harvested along with paddy harvesting. This makes very difficult in marketing the fish.
- Most of paddy fields are irrigated by common water channels and it is very difficult to check for water quality/pesticides.
- Growing fish in paddy fields may require more investment then rice culture alone.
- Rice and fish farming requires more labour then paddy cultivation alone.

Polyculture

Polyculture is the practice of culturing more than one species of aquatic organism in the same pond. The motivating principle is that fish production in ponds may be maximized by raising a combination of species having different food habits. The mixture of fish gives better utilization of available natural food produced in a pond. Polyculture began in China more than 1000 years ago. The practice has spread throughout Southeast Asia, and into other parts of the world.

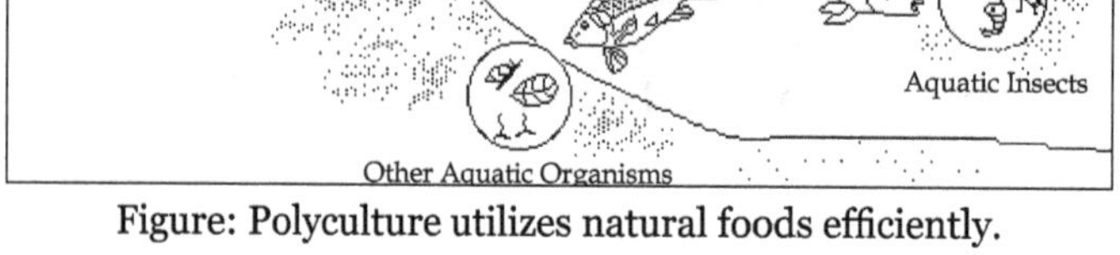

Figure: Polyculture utilizes natural foods efficiently.

Ponds that have been enriched through chemical fertilization, manuring or feeding practices contain abundant natural fish food organisms living at different depths and locations in the water column. Most fish feed predominantly on selected groups of these organisms. Polyculture should combine fish having different feeding habits in proportions that effectively utilize these natural foods. As a result, higher yields are obtained. Efficient polyculture systems in tropical climates may produce up to 8,000 kg of fish per hectare per year.

Fish used in Polyculture

Combinations of three Chinese carps (bighead, silver and grass carp) and the common carp are most common in polyculture. Other species may also be used. While fish may be grouped into broad categories based on their feeding habits, some overlap does occur. Descriptions of the feeding habit categories and examples of fish from each category follow:

Plankton Feeders

Plankton is normally the most plentiful food in a pond so it is very important to include a plankton feeding fish in a polyculture system. This group of fish feeds on the tiny, free floating plants (phytoplankton) and animals (zooplankton) which multiply abundantly in fertilized ponds. Two fish typical of this group are the silver carp, Hypophthalmichthys molitrix, and the bighead carp, Aristichthys nobilis. Other plankton feeders include:

Country/Region	Local Name	Scientific Name
China	ma lang yu ca choi striped mullet	Squaliobarbus curriculus Labeo collaris Mugil cephalus
India	Catla fringe lipped carp white carp Cauvery carp Bata	Catla catla Labeo fimbriatus Cirrhinus cirrhosa Labeo kontius Labeo bata
Indochina	sandkhol carp	Thynnichthys sandkhol
Vietnam	ca duong	Hypophthalmichthys harmandi
Worldwide	blue tilapia blue tilapia	Oreochromis aureus Oreochromis niloticus

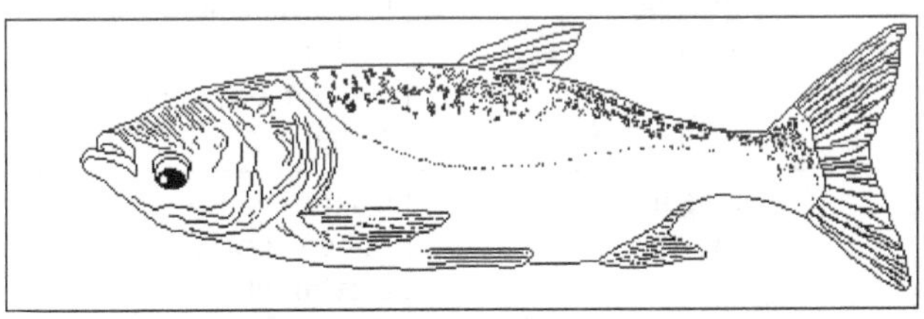

Figure: Silver carp feed primarily on phytoplankton.

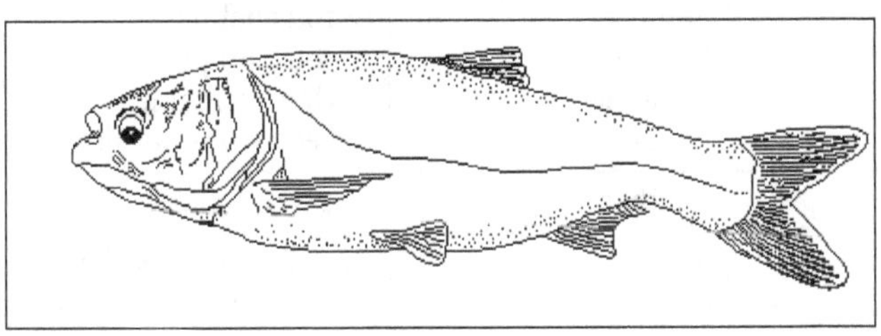

Figure: Bighead carp feed primarily on zooplankton.

Herbivores

This group of fish feeds on aquatic vegetation. The grass carp, Ctenopharyngodon idella, is most noted for this behavior and is stocked in ponds for the purpose of weed control. Other herbivores include:

Country/Region	Local Name	Scientific Name
Africa	Tilapia	Tilapia rendalli
India	Rohu Cauvery carp reba	Labeo rohita Labeo kontius Cirrhinus reba
Indochina	ca ven	Megalobrama bramula
Indonesia	giant gourami	Osphronemus goramy
SE Asia	Tawes	Puntius gonionotus
World wide	Zillis tilapia	Tilapia zillii

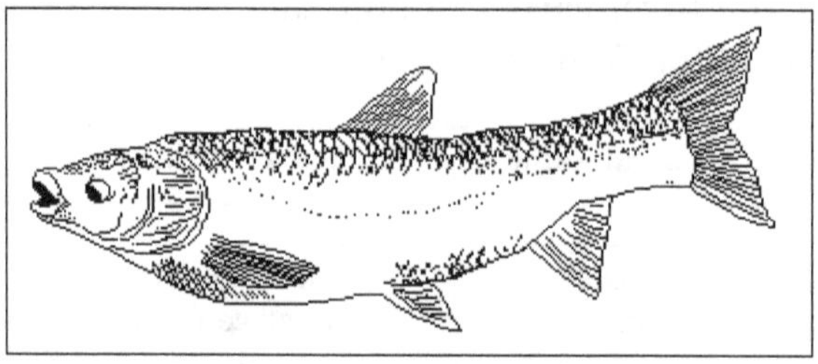

Figure: Grass carp feed on aquatic vegetation.

Bottom Feeders

Fish in this group feeds primarily at the pond bottom. They consume a variety of decaying organic matter, aquatic organisms such as clams, insects, worms, snails, and bacteria living in or on the sediments. The common carp, Cyprinus carpio, is well noted for this behavior. Other bottom feeders include:

Country/Region	Local Name	Scientific Name
China	Mud carp Black carp Cha cham Striped mullet	Cirrhinus molitorella Mylopharyngodon piceus Mylopharyngodon aethiops Mugil cephalus
India	Mrigal Cauvery carp Reba Nagendram fish	Cirrhinus mrigal Labeo kontius Cirrhinus reba Oteochilus thomassi
SE Asia	Belinka Lampai mata merah tambra	Barbus belinka Barbus lampai Barbus orphoides Labeobarbus tambroides
Taiwan	Milkfish	Chanos chanos
World wide	nile tilapia blue tilapia black tilapia	Oreochromis niloticus Oreochromis aureus Oreochromis mossambicus

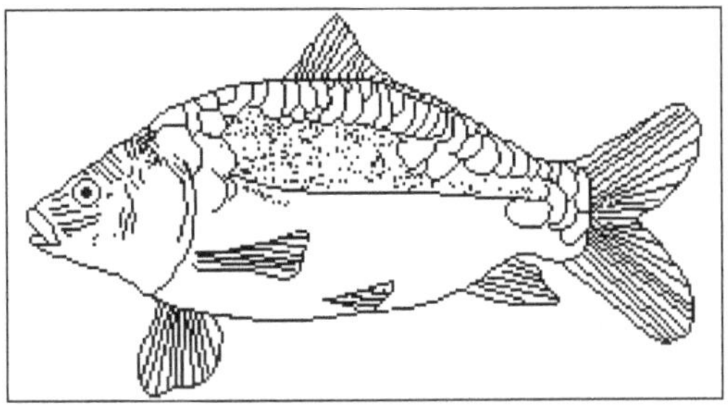

Figure: Common carp are bottom feeders.

Piscivorous Fish

These predatory fish feed on other fish, and must consume about 5 to 7 g of prey in order to grow 1 g. They are frequently stocked in ponds to control unwanted reproduction, particularly in tilapia, and other fish that enter the pond with the water supply and compete for food with the stocked fish. Commonly used predator fish include the sea bass, Lates spp.; catfish, Clarius spp. and Silurus spp.; snakeheads, Ophicephalus spp.; cichlids, Cichla spp.; Hemichromis fasciatus and Cichlasoma managuense; knife fish, Notopterus spp.; and largemouth bass Micropterus salmoides.

Table: Various predator fishes used in polyculture.

Lates	Ophicephalus	Cichla	Hemichromis fasciatus	Clarius
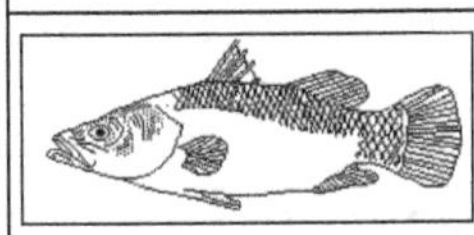	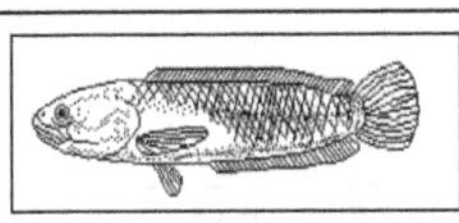			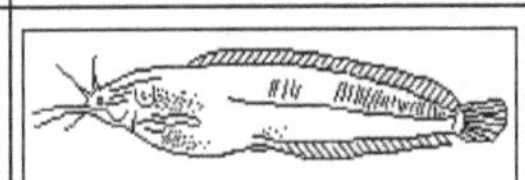

Adding predator fish to a polyculture system increases the average weight of prey species. It is most efficient to use a predator fish that consumes small prey. This prevents the prey from growing large enough to compete for food with larger fish of its species. Use of predator fish in polyculture systems is experimental in most areas of the world. In small ponds, it is almost impossible to stock the exact number of predator fish to achieve the same predator/prey balance occurring in nature. In small scale aquaculture, predator fish are usually stocked at rates of 5 to 20 fish per 100 m2 of pond surface area to completely control reproduction of the prey species.

Factors affecting Species Selection and Stocking Rates

- Water temperature: Many fish cannot survive or grow well in cold water. Systems using cold-tolerant fish such as common carp and Chinese carps must be used if temperatures drop below 18°C.
- Market value of fish: The market price and demand should be considered before a fish species is chosen for culture. When two or more fish can fill the same feeding niche in a pond, the choice should be based on which will maximize economic returns to the farmer.
- Pond fertilization practices: Most polyculture systems are based on fertilization. Manures and chemical fertilizers increase production of natural fish food organisms in ponds. Thus,

more food is made available to fish. Fertilized ponds may be stocked at higher rates than unfertilized ponds.

- Feeding habits of fish: Supplemental feeds are commonly given to fish. Manure may serve as a food source for some fish by supplementing the nutrition available from natural food organisms in the pond. A wide variety of agricultural by-products may serve as supplemental feed. When fish are fed, ponds can be stocked at higher rates. Stocking bottom feeding fish such as common carp prevents sinking foods from being wasted.
- Tolerance to pond conditions: Polyculture ponds are usually heavily fertilized or manured. This practice may cause low oxygen levels and other conditions in the water that are stressful to fish.
- Potential of uncontrolled spawning in grow-out ponds: Certain fish, like tilapia, reproduce easily in ponds. Tilapia may become so overpopulated that their growth stops and they become stunted. Predator fish are often stocked in tilapia ponds to control reproduction.

Stocking Rates for Polyculture Systems

Examples of stocking rates for polyculture systems used in various countries are presented in table, and are intended as general guidelines. Modification to suit conditions in other locations may be necessary.

Table: Number of fish stocked per 100 m^2 of pond surface area in polyculture systems used in various countries.

Species	CHINA	INDIA	MALAYSIA	THAILAND				PANAMA	SIERRA LEONE
Bighead carp	1	-	1	3	-	-	-	10	-
Silver carp	12	-	1	3	-	-	-	-	-
Grass carp	2	-	3	3	-	-	-	-	-
Common carp	7	-	1	6	32	31	-	10	-
Tawes	-	-	-	63	63	63	-	-	-
Rohu	-	38	-	6	-	-	-	-	-
Mrigal	-	6	-	-	-	-	-	-	-
Catla	-	19	-	-	-	-	-	-	-
Tilapia	-	-	-	63	63	63	125	100	160
Ophicephalus	-	-	-	-	-	-	3	0	-
Cichlasoma	-	-	-	-	-	-	-	20	-
Notopterus	-	-	-	-	-	-	-	-	16

Potential Problems in Polyculture

Polyculture is an effective way to maximize benefit from available natural food in a pond. But, pond management becomes more difficult when stocking fish species having specialized feeding habits in the same pond because good fertilization and feeding practices must be followed. If inadequate fingerling supply severely limits the choice of species available for polyculture, at least one species should have general rather than specialized feeding behavior. This will allow more of the available natural food to be utilized.

References

- Fuel-efficiency, algaculture, biofuels: howstuffworks.com, Retrieved 03, Jan 2020
- What-is-the-environmental-impact-of-aquaculture: aquaculturealliance.org, Retrieved 14, Jan 2020
- Fish-pond-culture: fao.org, Retrieved 12, Feb 2020
- Fish-Farming-in-Recirculating-Aquaculture-Systems-RAS: fisheries.tamu.edu, Retrieved 25, Feb 2020
- Rice-fish-systems-fact, crop-establishment: knowledgebank.irri.org, Retrieved 04, March 2020
- Rice-fish-farming: agrifarming.in, Retrieved 15, March 2020
- Integrated-rice-and-aquaculture-farming, aquaculture-plants-and-invertebrates: intechopen.com, Retrieved 26, March 2020

3

Types of Fish Farming and Fisheries

Fish farming is a form of aquaculture in which fishes are raised in tanks or other similar enclosures for personal use or profit. Extensive aquaculture and intensive aquaculture are two basic types of fish farming. The occupation or enterprise of undertaking such an activity is known as a fishery. This chapter deals with fish farming and fisheries and the various types, practices and the issues plaguing them in a detailed manner for the reader's benefit.

Fish Farming

Fish farming refers to the commercial production of fish in an enclosure or, when located in a body of freshwater or marine water, in an area that is penned off from the surrounding water by cages or open nets.

A fish farm is similar to a fish hatchery in that both can contain 500,000 and more fish. But, a fish hatchery is designed to raise the fish only to a young age before they are released into the wild, usually to bolster the numbers of that species. In contrast, a fish farm is designed to raise the fish until they are a size and age that makes them the best commercial value. The fish are ultimately retrieved and sold, typically as whole or processed food.

Fish farming is the most common form of aquaculture, and commonly involves trout, salmon, tilapia, cod, carp, and catfish. For a species such as cod, whose numbers in the Grand Banks fishery off the east coast of the Canadian Maritime Provinces plummeted to near zero in the 1970s due to overfishing, and as of 2008 have yet to recover, the cod available from fish farming represents almost the sole source of the fish in North American markets.

The example of cod is cited as one of the advantages of fish farming. Raising fish under more controlled conditions that are possible in the wild avoids the problem of overfishing. As well, because an operation takes up relatively little space, feeding and care of the fish can be done under more controlled conditions, which is an economic advantage to those who own and run the facility.

However, fish farming is a controversial practice. For example, on the Canadian west coast, the farming of salmon typically uses species normally found in the Atlantic Ocean. The escape of fish to the wild does occur, and has created concern that the presence of the species in an environment that is unnatural to them could upset the marine ecology. Other concerns of fish farming are the overcrowding of fish, which can make them more susceptible to disease such as sea lice, and the use of antibiotics, which can also be released into the natural environment.

Fish farming is an ancient practice, dating back to about 2500 BC in China, when carp were raised in ponds and in artificial lakes created by receding floodwaters. Some of the motivations for fish farming in ancient China are shared with fish farm owners and operators in 2008. These include maximizing the food available from the resource; reducing the energy needed to search for, gather, and transport the food; making food production more predictable and less likely to be influenced by weather, predators, or other factors; and ensuring that the quality of the resource remains acceptable over time.

Evidence of fish farming also dates back at least 1,000 years in Hawaii, when rocks were added to existing reefs to create an artificial pond. The spaces between the rocks were large enough to let seawater circulate in and out, allowing nutrients to circulate in and waste out, but were too small to allow the fish to escape. The open net design of present-day fish farms follows this example.

Fish farming became more prevalent in Europe in the fifteenth century. The first known fish hatchery constructed in North America was built in the Canadian province of Newfoundland in 1889.

In the 1960s, fish farming expanded worldwide as some commercial fish stocks became less plentiful and a growing global population increased the demand for fish. As with factory farms—the landlocked facilities in which huge numbers of poultry and livestock are raised—economic incentives were provided to encourage the establishment of freshwater and marine water fish farms. In addition, corporations involved in the sales of fresh and processed fish products began to expand into fish farming as a way of ensuring supply, expanding their market, and trimming costs.

As practiced in China thousands of years ago, fish farming was efficient and sustainable. The numbers of fish were suitable for the space available and the population was managed so that the numbers of fish ready for harvesting did not decline over time. When done in such a sustainable way, fish farming can be a good strategy to supplement or even replace the fish caught from the wild.

However, the confinement of large numbers of fish in a small area can create problems. In an enclosed pond, one problem can be the accumulation of waste products. Aside from making the water less hospitable for the fish, the waste material can act as a food source for microorganisms known as algae. Combined with suitable water temperature and sunlight, the presence of the food source can lead to the rapid increase in the number of algae termed as algal blooms. The number of algae in blooms that occur in the open ocean can be so large that the growth is visible from orbiting satellites. In a confined pond, an algal bloom can use up much of the oxygen in the water, leading to the death of the fish.

When a fish farm is done in tanks (closed circulation type), it is essential to keep the water well oxygenated and to remove wastes. Bubbling air into the water as is done with a home aquarium can be one means of oxygenation. Alternatively, water can be cascaded from tank to tank in the fish farm complex, with oxygen being supplied as the water tumbles between tanks. Waste removal is typically done by continual removal of the water and passing it through a filter before returning the water to the tank. Regular monitoring of the tank water is necessary to ensure that other parameters such as pH are maintained at an optimal level.

Fish farming that is done in large bodies of freshwater or in the ocean avoids the potential problems associated with the closed circulation system. Typically, the fish are housed in a series of pens, which consist of mesh nets attached to rigid supports. The entire structure floats on the surface.

Impacts and Issues

Fish farming has become a very contentious practice, for a number of environmental reasons and for the adverse health effects it has on the farmed fish and possibly other species, including humans.

In a fish farm, the concentration of fish far exceeds that found in schools of fish in the wild—50,000 or more fish in an area of several acres in volume—with the possible exception of the spawning runs of west coast salmon. These crowded conditions reduce the free-swimming volume of each fish to about that of the average household bathtub. In such crowded conditions, the fish bump and rub against each other in the boundaries of the pens, which can produce cuts and scrapes. This increases the likelihood of infection and the development of diseases.

Species of sea lice that parasitize Coho and Atlantic salmon are especially troublesome. The sea lice attach to the fish and feed on tissue, which creates lesions and causes fluid loss from the affected fish. The confined fish become ill and can die. In addition, the sea lice can spread to wild salmon in the seas around fish farms when farmed salmon escape from the confinement, and also when the lice are washed away from the fish farm into the surrounding water. A 2001 survey of wild juvenile salmon migrating past fish farms in British Columbia found many more sea lice on the juveniles that had passed the farms than on those who had not yet passed by the facilities.

The escape of fish from fish farms is not a trivial and isolated event. Rips and breaks in the pen material and buffeting of the pens by storm-driven waves can lead to the escape of fish. In some cases, pens are designed with a net lid to reduce this possibility. Sometimes only a few fish escape. But mass escapes have occurred. For example, in January 2002, over 8,000 fish escaped from a fish farm in Clayoquot Sound, British Columbia. Worldwide in 2004, an estimated 2 million farmed fish escaped to the wild.

Once in the wild, the farmed fish have the potential to transfer disease to the wild population. An article in a December 2007 issue of Science documented declines in the population of wild Pacific salmon related to their decimation from sea lice transferred from farm populations of Atlantic salmon. The situation is so dire that the natural population could be reduced by 99% by 2015, which would be an economic disaster for the traditional salmon fishery and those employed by the fishery.

Antibiotics can be supplied to the food in an effort to control infections. As with land-bound factory farms, this practice encourages the development of antibiotic resistance among the surviving bacteria. These hardier varieties of the bacteria may pose a health hazard not only to the farmed fish, but to wild fish populations and to humans.

A fish farm releases a great amount of untreated sewage to the surrounding water. A study done in Clayoquot Sound calculated that the 700,000 fish housed in the facility that is the size of three football fields generate the daily equivalent amount of sewage produced by 150,000 people.

Extensive Aquaculture

Extensive aquaculture includes systems of culture and rearing in which human intervention is concentrated on the reproduction of the stock, in addition to capture. As compared to fishing, these systems make it possible to selectively increase the production of the species most useful for man's food (salmon, oysters, etc.), industry (algoculture), or pleasure (sport fishing, pearl culture, etc.). They use the planktonic or benthic production of the water masses, thus economising on the fertilisation of the breeding environment (semi-intensive systems), or the fattening of the stock (intensive systems).

Extensive practices are not new. In Europe, the fattening of mussels on stakes dates back to the 14th century. The value of capturing spat as a remedy for the overexploitation of natural oyster beds was realised as early as the second half of the 19th century. The value of transplanting young plaice from their nurseries in the south of the North Sea to less densely populated lagoon or sea fattening zones, at the beginning of this century. The mastery of artificial reproduction stimulated enthusiasm for sea ranching: the first releases of young salmon from hatcheries date from the middle of the last century, the first cod hatchery was built in Norway in 1884, and the first lobster hatchery in 1891 on the Atlantic coast of Canada.

Knowledge about the actual production of several extensive systems is still limited. While the production of sedentary species (shellfish and algae), whose stocks are geographically distinct from the natural beds and generally owned by an individual or a corporate body, is fairly well known, the culture of vagile species (mainly fish and crustaceans) whose stocks are exploited collectively at the same time as wild stocks and whose captures are not distinguished from those of the natural populations, is not at all well known. This uncertainty has led the FAO to exclude from its aquacultural statistics the production of collectively exploited stocks. This decision leads to the inclusion in the fisheries statistics of production resulting from the release of young fish into water courses, lakes, coastal waters, and the ocean, even though this production is very significant for several species, in particular salmon.

Algoculture and shellfish culture, with a production of 7,7 and 7,3 million tonnes respectively in 1996, based almost entirely on the extensive modes, represent half of world aquacultural production as defined by the FAO. European shellfish production exceeds 750,000 tonnes. In Japan, the production of chum salmon by ranching has increased tenfold over the past thirty years, to exceed 200,000 tonnes today. In 1994, the number of migrating fish in the rivers of Hokkaido was ten times the previous maximum observed in the entire history of fishing.

From the ecological standpoint, the potential for the expansion of extensive aquaculture appears considerable. Theories on the reproduction strategies of aquatic populations, the amplitude of the natural fluctuations of wild populations, and the productions achieved in certain extensive farming systems indicate that aquatic ecosystem could support stocks far bigger than the average biomass of the wild stocks. Furthermore, since the volume of the marine ecosystems exploited by these methods is still relatively small, it is understandable that this mode of production should stimulate research interest. The number of marine species studied rose from 150 in 1984 to 250 in 1994.

The results have been disappointing however. In recent years, few R&D programmes have led to any significant production. Shellfish culture is experiencing overstocking problems which are resulting in the deterioration of the technical and economic performances of the enterprises concerned. In the effort to reduce these imbalances and manage the conflicts which result from the competing uses of coastal ecosystems, traditional methods of regulation have proved ineffective.

The prospects for extensive aquaculture can be evaluated with reference to the state of the sector. Since the 70s, the annual rate of increase in world halieutic production has constantly been falling - in value terms even more rapidly than in tonnage -, to become negative in the course of the this decade. The reason for this decline is the exhaustion of wild stocks and the ineffectiveness of the traditional methods of regulating access to fisheries whose possibilities for expansion are exhausted. The aquaculture development policies introduced at the end of the 60s to take over from conventional fishing started to bear fruit during the 80s. For the past fifteen years world aquacultural production has been growing at an annual rate of 8.5 %. It now accounts for over a quarter of total production. Progress remains unevenly distributed however in terms of cultivation systems, production regions, products and markets. Despite the spectacular success of salmon farming in northern Europe or prawn culture in the tropics, traditional systems (small shellfish farms, fresh-water fish farms, and algoculture, mainly in Asia) remain the main beneficiaries of the new expansion. Due to the stagnation of fishing catches, European countries as a whole are faced with both a quantitative and qualitative shortage of sea products. In 1996, they imported over US$ 21 billion of sea products, and the deficit amounted to US$ 6 billion. For the European Union alone, it was US$ 8 billion - and respectively US$ 16 billion and US$ 4 billion for Japan and the USA. Only Norway, Iceland, Russia, Denmark, Ireland and Ukraine are, in descending order, net exporters in value terms. As in most of the rich countries, the shortage was mainly in high quality products (noble fish, crustaceans and shellfish, fresh, filets and frozen). In these conditions, public policies should continue to support the expansion of aquaculture and the diversification of production systems, with priority being given to those concerned with species of high commercial value.

The marine environment is well suited for the analysis of the implications and possibilities of extensive aquaculture. Because of their large dimensions, and because, even in coastal areas, the seas and oceans suffer on average less pressures than inland waters, marine ecosystems offer better prospects. The scale of the oceanic masses reveals better the problems that the fluidity of the milieu and the mobility of the stocks represent for their domestication. Any advance in the law of the sea sheds better light on the role of the institutions in the exploitation of natural resources and the rationalisation of their uses.

Main Extensive Systems

Process of Domestication

Technical intensification makes it possible to exceed the productivity limits of the wild stocks. The process consists of extending the technical controls of physiological functions such as reproduction, particular stages of the life cycle (eggs and larvae, juveniles, adults, reproducers), as well as their cultivation milieus. This domestication takes place through discrete jumps: between fishing and completely controlled cultivation there is a range of systems that can be classified in four groups: (i) those which concern essentially reproduction, (ii) those which also intervene in the fertility of the cultivation milieu, (iii) those in which the stock is also fed, and (iv) lastly, those in which the wastes are also recycled. Not all of the technically conceivable systems are viable, and the most intensive are not necessarily the most efficient.

While technical intensification reduces the role of the natural resource, it does not eliminate the dependence of production vis-à-vis the environment. It changes only the nature of the constraining factors and the limits that they impose on production. In fishing, it is the tertiary production - that of the halieutic populations - which is the limiting factor. In aquacultural systems, the stock, which is artificially reproduced, is no longer limiting. It no longer constitutes the natural resource. Appropriated, it becomes assimilable to capital. In extensive cultivation it is the primary and secondary productions (plankton and benthos) serving as food for the stocks which limit production. In intensive systems where the food is added it is the wastes (unconsumed food, faeces, antibiotics and medicines, etc.), as well as escapes of individuals who may modify the genetic heritage of the wild populations, that need to be adjusted to the acceptance capacity of the aquatic ecosystem. Throughout the process of domestication, the capital and labour intensity of the operation increase at the same time as the technical intensification. The increase in productivity (per unit of area) thus obtained is accompanied by an increase in the pressure on the environment.

In the course of this process the economic and social organisation of the operation also changes. As the technical controls diversify, the rules which define and support their exercise - and in particular the regimes governing the ownership of the natural resource - change. Mazoyer and Roudart thus describe the links which have developed in agriculture between the domestication of the natural resources, the intensification of their exploitation, and the land ownership regime: 'It emerges from this long history that, since the Neolithic age, the "ownership" of the land has gradually extended to the different categories of land as they have become artificialised: built on land first of all, then gardens and enclosures enriched and cultivated each year, cleared land bearing crops, developed hay meadows, land cultivated between two uncultivated or fallow periods, land cultivated continuously, improved pastures, developed and maintained forests. The ancient rights of common use (hunting, gathering, gleaning, use of the common, right to collect wood), for their part, have always concerned land where, without any particular work, wood, natural grasses, the wild grasses of fallow land, game, etc. continued to develop spontaneously'. The history of shellfish culture in Yerseke in the Netherlands shows that the processes in aquaculture do not differ fundamentally from those observed in agriculture. In the course of the process of technical intensification, the structures of the production, exploitation and regulation systems were transformed in stages:

- Before 1860: collective exploitation of natural beds by small local fishermen; regulation of access by the administration;
- From 1860 to 1933: with differences between oyster culture and mussel culture, transition from fishing to extensive aquaculture, with the public auction of individual concessions in the public maritime field, and privatisation of the stocks; application by shellfish culture cooperatives of individual production quotas aimed at regulating exports;
- 1933 - 1967: intensification of production; epizootic diseases; administrative regulation of production and markets;
- After 1967: vertical integration of enterprises, which become capitalistic; reduction of public intervention.

Description of Extensive Aquaculture

Extensive aquaculture is characterised by a production function in which human intervention is concentrated on the reproduction of the stock. This intervention may take two forms:

- Through the physical creation of particular habitats: seasonal closing of lagoons to hold the juveniles before their migration to the sea (valliculture, which includes other interventions such as the regulation of the water circulation, the elimination of undesirable species, or the provision of additional food, etc.), creation of fry ponds and artificial nurseries to support the reproduction of natural populations (salmon), capture of natural spat by providing substrata for the collection of larvae (shellfish culture), creation of niches for the juveniles of certain species when the density of the former is a limiting factor, etc.

- By releasing hatchery fry into the natural milieu.

Representation of Extensive Aquacultural Systems

The first set of examples shows that artificial reproduction is not essential to start a new system. On the other hand, the capacity to act on the reproduction of a population through forcing its recruitment is.

Extensive systems aim at three objectives:

- Reconstitution by restocking of natural populations reduced by the deterioration of their environment, in particular the degradation of habitats which are critical for reproduction (pollution, dams on water courses which impede the migration of reproducers, disappearance or silting up of the fry ponds and nurseries of anadromous fish, etc.);
- Acclimation of exotic species or populations in order to create new stocks more valuable than the native stocks;
- Simple enlargement of a stock by releasing young fish into the natural milieu, often resulting in overpopulation.

As the stock feeds on the natural milieu, extensive fish farming is sometimes qualified as 'production', as opposed to intensive fish farming, qualified as 'processing' in which artificial feeding is provided. In the sea, semi-intensive systems in which at least part of the food is produced in the cultivation milieu do not exist (outside lagoons). This is not the case in agriculture or fresh-water fish culture. This difference is explained by the fluidity of the aquatic milieu and the scale of the water masses which limit the possibilities of fertilisation to units of sufficiently reduced dimensions - hence inland (ponds, lakes and lagoons) -, and also by the state of ownership regimes in the maritime field. In the sea, the ownership of stocks is limited virtually exclusively to sessile or sedentary species (shellfish, algae), and vagile species (fish) in captivity (fish culture in cages or basins). The importance of intensive cultivation in the marine field is perhaps due as much to the need for the appropriation of the stock as to the enhanced zootechnical performance.

Extensive systems which aim at fattening naturally result in overpopulation (as compared with the biomass observed under natural conditions). The increase in biomass – and hence finally in the total production of the stock - is limited by the trophic capacity of the ecosystem used as the cultivation milieu. The growing tension between the biomass in expansion and the limited productivity of the cultivation milieu results in a regular decline in yields (decline in growth performance, increased mortality and increased risk of epizootic diseases).

Extensive cultivation does not modify the environment very much. However, the employment of selected strains for the purposes of cultivation or exotic strains for the rearing of young fish risks impairing the genetic heritage of the wild populations living in the same ecosystems.

Recapture is facilitated when the stocks are made up of sessile or sedentary species, or when fishing is carried out close to the coast or in rivers where the stock concentrates periodically (salmon during their reproduction, for example).

Like halieutic stocks and intensive cultivation, extensive aquaculture is subject to the effects of pollution by industrial, agricultural and domestic wastes. The fluidity and density of aquatic milieus makes them more sensitive than land and atmospheric milieus. Aquatic milieus are in fact better carriers than land and atmospheric milieus, and leaching and transport by water course networks concentrate the greater part of the residues from human activities in the catchment areas in the rivers and coastal areas. On the world scale it is estimated that some 70 % pollution ends up in the sea.

Physical Development of Habitats

A halieutic population can be supported by the development of habitats critical for its reproduction. This is the case, for example, with ladders to enable salmon to get past dams on the rivers they use for their reproduction migration, arrangements to conserve the flow and quality of river water, or the maintenance of their fry ponds.

Artificial reefs are the most common example of physical measures in the marine field. It is in Japan, where it is planned that they will cover 12 % of the coastal areas by the year 2000, and along the Mediterranean coasts in Europe, that these arrangements are the most developed.

Their promoters attribute the following virtues to artificial reefs:

- Concentration of the fish by tropism (provision of food and protection against predators);
- Increase in biotic diversity, for example when the reefs are installed in sedimentary zones poor in sessile and sedentary species (algae, shellfish, urchins, crustaceans, rocky ground fish, etc.);
- Increase in the biomass, effectively observed in the case of sedentary and sessile species, but not so obvious in the case of vagile species (fish);
- Physical elimination of tackle such as trawl nets.

Depending on the objective, the species it is desired to support, and the fishing methods it is intended to favour, the materials and architecture of artificial habitats vary considerably.

In the creation of artificial reefs, the exclusion of non-indigenous species by residents preoccupied by the rarefaction of natural resources in the sectors that they are accustomed to frequent is often an important motive, even though it is not always expressed. This monopolising of space may facilitate initiatives in favour of conservation. In the USA, for example, artificial reefs are used to increase the floral and faunal richness and promote the development of ecological tourism. The example of floating rafts and other arrangements to concentrate fish shows that these constructions can effectively attract fish. Through bringing the fishing grounds closer, stabilising the presence of the fish and facilitating the spatial allocation of fishing sites, artificial reefs can aid fishing and its regulation. But the same results can also be achieved by other methods, possibly less costly (floating rafts), or more direct (shellfish culture).

The effect of artificial reefs on the production of halieutic stocks remains to be demonstrated. Up to now, the observations have virtually exclusively been concerned with spot densities of adults, whereas to demonstrate the impact of the reefs on halieutic productivity it would be necessary to be able to measure their effect on the absolute abundance of the populations. This implies studying the process of regulation, on the scale of the population, and at the level of reproduction and the early stages where these processes act. While such studies may be complex and costly, they are nevertheless realisable. Only such investigations would make it possible to establish whether artificial reefs are the best way of achieving the objectives assigned to projects for the implantation of reefs and, if this is the case, the conditions that must be fulfilled (nature, localisation, extension) to obtain the best impact. Artificial reefs, for example, are likely to effectively support the recruitment of halieutic stocks provided that they cover a sufficient proportion of the coastal nurseries. On the other hand, through acting on the effects (stocks) and not on the causes (regulation of access) of overfishing, they can at best retard, but not prevent, overexploitation.

Shellfish Culture

In shellfish culture, the forcing of recruitment is achieved in different ways:

- By Simply taking juveniles from the natural beds and replacing them on the bottom or in the water on artificial supports (hurdles or nets) for fattening, - i.e. in milieus that the larvae cannot colonise under natural conditions; this technique is widely used in mussel culture;
- By capturing spat on artificial supports placed in the natural milieu; the larvae which attach themselves on the collectors come from natural beds and, above all, from cultivated stocks once the system is well established; as with the preceding technique, the spat is then transferred to the fattening milieu (on the bottom or in suspension); this technique is that most generally used in oyster culture;
- By the production of juveniles in a hatchery (scallops, oysters, abalone, Philippines clams (Ruditappes philippinarum), etc.).

The more or less complete mastery of recruitment has permitted three major advances:

- Creation of stocks whose biomass may considerably exceed that of the natural populations; thus, in the Marennes-Oléron basin (France), the oyster biomass now regularly reaches 200 000 tonnes, or several times the highest ever reached by the natural beds;
- Freedom from the climatic hazards of natural recruitment and their effects on annual production; again for the Marennes-Oléron basin, the capture of spat has been zero for only two years out of 45 years of observation for the Portuguese oyster, and low or zero only three years out of thirteen for the Japanese oyster (Héral et al., 1986);
- The extension of production to sectors where the species does not reproduce; this is how in France, the cultivation of the Japanese oyster north of the Loire is achieved thanks to spat captured south of this river.

Thus, a rustic production chain, not initially requiring the complete mastery of the biological cycle, has turned out to be economically efficient and well adapted to the exploitation of particularly productive coastal ecosystems. This development has taken advantage of three particularities of shellfish ecology:

- Their sedentariness which facilitates the confinement and appropriation of the stock;
- Feeding by filtration which makes it possible to exploit the most productive lower levels of the food chain and to achieve a high concentration of cultivation;
- Their good tolerance of climatic variations.

But the take-off of the system was achieved only because these technical advances were complemented by institutional innovations:

- Privatisation Of The Stocks: Made Physically Possible By The Sedentariness Of The Species, The Possession of the stocks has been formalised by the attribution to oyster growers of private concessions on the public maritime domain; having the necessary guarantees for their investments, the growers have been able to invest in the reproduction of the stock;
- Intervention by the public authorities in the conservation of the quality of the oyster farming ecosystem, the control of the health of the stocks and the health quality of the products.

Shellfish culture has been developed above all in Asia (China, Korea, Japan, etc.) and Europe. In France, for example, the shellfish culture turnover amounts to over one-third that of fishing and employment is one and a half times that of fishing. The big differences seen in European production, both between countries and between species, suggest that the constraints which impede the development of shellfish culture are social and institutional rather than ecological or technical. There should therefore be possibilities for expansion in countries where production is still modest.

90% of European production is accounted for by oysters and mussels. While both oyster and mussel growing have been modernised in recent years, Europe has not achieved any successes comparable to that of Asia with the development of the cultivation of the Japanese scallop (Pecten yessoensis). Mastered in Japan at the beginning of the 50s, production reached a million tonnes in China and 265 000 tonnes in Japan in 1996. There are nevertheless prospects for diversification in Europe. The cultivation techniques of several species have been mastered, but the take-off of commercial production comes up against various difficulties: ecological and economic for the abalone (Haliotis tuberculatus), epizootic (brown ring disease) for the Philippines clam (Ruditappes philippinarum), and legal for the scallop.

Sea Ranching

While, as with the sedentary species, the fattening of vagile species is possible in the marine milieu, their mobility seriously complicates the appropriation of the stocks and the captures. The systematic marking of hatchery products is made difficult by the small size of the individuals. Above all, it is difficult to ensure that the ownership of the stocks is respected at sea because of the remoteness of the fishing areas and of the low density of boats on them, as well as the transitory nature of the catches. This is why the ownership of stocks of vagile species is practically never acknowledged at sea. Contrary to what happens with livestock in agriculture, fish that escape from private farms become public property.

For certain species these difficulties have been partially overcome by the exploitation of a trait common to many aquatic populations, the phenomenon of homing. Through choosing species which congregate to spawn in coastal or inland waters where access to fishing is restricted and easier to control, enough of the stock can be recaptured to cover the cost of rearing the fry. Sea ranching is based on this strategy. Technical success is assured if the weight gain in the individuals recaptured during the fattening phase exceeds the combined losses due to natural mortality, straying – not all individuals return to the point where they are released -, and catches by third party fleets. When these conditions are fulfilled sea ranching offers the possibility of harnessing the unexploited primary and secondary productivities of vast maritime areas.

The risk of interception by third party fleets increases, on the other hand, with the area and duration of the freedom phase. The salmon ranching programmes of the Baltic countries (mainly Sweden and Finland) have been so successful technically, for example, that they have given rise to the development of plurinational fishing at sea. The captures have reached such a level (some 3,000 tonnes, or 80 to 90 % of the annual recruitment) that they are now threatening the wild stocks fished at the same time. Through ensuring the profitability of fishing, a rearing programme intended to conserve the natural populations decimated by hydraulic schemes and river-borne pollution is now threatening their survival. The answer would be to prohibit fishing at sea. This is the solution adopted in the North Pacific where the states have agreed to prohibit salmon fishing on the high seas, or in Iceland where salmon fishing has been prohibited in coastal waters for the past 60 years.

The system was first developed for salmon. For these species in fact, ranching offered a solution to the crucial problem of the conservation of populations threatened by the degradation of their inland water habitat on which they depend for their reproduction and in which they are particularly vulnerable. Salmon exhibit a number of traits that favour this type of enterprise:

- they reproduce in very localised fry beds in water courses, where their recapture is easy, after a phase of oceanic fattening in the course of which the individuals acquire over 90% of their final weight;
- they have a very marked homing behaviour, with a generally low rate of straying;
- the particularities of their demographic dynamic (the cohorts reach their maximum biomass at the moment of reproduction, after which they all die - Atlantic salmon - or most of them die - Pacific species) theoretically make it possible, through short-circuiting the freshwater phase, to compensate for the problems with the conservation of inland waters and the halieutic stocks; again in theory, the regulation of their fishing could be replaced by simple management of the annual releases;
- artificial reproduction is easy to achieve (salmon are one of the small number of aquatic species which lay large eggs), and was mastered very early;
- the species is of high value for human food, or even more for sport fishing.

For a long time the impact of the experiments carried out in the Atlantic and North Pacific was nil or not proven. It was not until the 60s that, simultaneously in Sweden and the USA, the technical viability of the method was able to be established. Since then, advances in knowledge have revealed the critical importance of factors which had been neglected for a long time, such as the marking and identification of hatchery individuals in catches for the evaluation of survival rates and conditions, or the risk of genetic mixing or pathological contamination with wild populations, etc. Since then the list of first technical and then economic successes has lengthened. In the literature, the turbot in Europe, and the gilt-head bream (Pagrus major), the flounder (Paralichthys olivaceus), the chum salmon (though perhaps no longer since the development of fish farming in cages has brought prices down) and the Japanese prawn (Penæus Japanicus) in Japan (Kitada, op. cit.), are cited among the species whose production by ranching is considered economically viable. On the other hand, in Europe, research into lobster (France, United Kingdom, Norway) and cod (Norwegian PUSH8 programme) reached negative conclusions.

The interest of sea ranching is not limited to the conservation of marine fauna. The technique may meet three objectives:

- Restocking for the conservation of natural populations. This is the most common case. Thus, when Sweden launched its hydroelectric programme, it imposed on its electricity company a compensatory fish rearing programme. The other Baltic countries subsequently introduced similar measures; in the second half of the 80s, these countries released over 6 million smolts each year. Whereas only 20 Swedish rivers out of 60 are now accessible to salmon, and natural recruitment has fallen by 90%, the resource is quantitatively conserved;
- Overpopulation to support commercial and recreational fishing: in Japan, the releases of chum salmon amount to 2 billion individuals a year, and the recapture rate increased from 2% in 1963 to 15% in 1995; on the East Coast of the Pacific, a decade ago salmon ranching provided 20 % of the total Canadian salmon catch, and one-third of the pink salmon caught in Alaska; but, as far as can be judged given the disparity of the statistics, on the world and European scales sea ranching production remains very much lower than that of shellfish culture or algoculture;
- Acclimation: there have been several instances of the successful introduction of non-indigenous species: chinook salmon in the rivers of the South Island of New Zealand; chinook and coho in Chile; coho in New Hampshire; but the results of these transplantations are often not very convincing, as in many cases the stocks can only maintained through repeated restocking with young fish, so that in the end these become simply natural milieu fattening programmes. They may be perfectly profitable however, as was concluded with the experiments carried out in the Kerguelen Islands in the Southern Ocean, for example.

The long history of transplantation of young plaice in the North Sea is worthy of mention. Launched in 1891 in Denmark and pursued for over a century, the programme has been extended to other species (cod, salmon, trout, whitefish, turbot and eel), and repeated in other countries (Germany, United Kingdom, Norway and Sweden). Although their technical efficiency has been demonstrated and economic profitability has been achieved in several configurations, these programmes have never given rise to significant commercial development. The reasons for this relative failure are political and financial. In Denmark, for example, the programme gives rise to disputes over the ownership of the stock. Fishermen of the coastal regions of the North Sea where the juveniles are fished claim they should conserve rights over the transplanted stock, just as the Limfjord fishermen, even though no professional fisherman contributes to the financing of the transplantation. Only the sports fishermen of the Limfjord pay a fee, and this represents only a small proportion of the total cost. In these circumstances, no significant private financing is going to take over from the public funding which for its part remains modest.

The organisation and management of ranching programmes differs according to the objectives. Programmes which are limited to the conservation of wild populations, without concern for profitability, are most often implemented by public bodies, even though the production of the young fish is sometimes subcontracted to private fish farms. The creation of significant stocks involves two questions: the reservation of their exploitation to particular groups (groups of national and local fishermen, professionals or anglers), and the contribution of the main beneficiaries of the catches to the financing of the breeding farms. The state of Alaska, for example, grants licences to private hatcheries for non-profit operations which are managed by local cooperatives and communities of native fishermen exploiting the stock. Fishermen can pay a fee for the financing of breeding programmes. This is the case in the United Kingdom, where the landowners finance fish rearing programmes and then rent the fishing rights on their reaches to anglers. In Japan, professional fishermen pay a contribution calculated on the basis of the sales price of the fish (3 to 5%) to finance coastal breeding farms, while access to the coastal fisheries requires belonging to a cooperative or a group of private fishermen. Certain countries (Iceland, Chile and, in the USA, the states of Alaska, Oregon, Washington, and California on an experimental basis, etc.) authorise the creation of farms whose aim is private production for profit. In most cases however, private ranching projects come up against the opposition of both professional and sport fishermen, as well as the general public, who object to the calling into question of public rights, the private exploitation of public resources, or the risk to professional fishing of a fall in fish prices.

Systems for Retaining Fish in Open Milieus or very Large Enclosures

In order to limit the losses due to the straying of fish and catches by third party fleets, attempts have been made to reduce the area of straying of the stock through conditioning juveniles before their release by emitting a sound signal during feeding. This training may facilitate the recapture of the fish. It is in Japan that experimentation is most advanced (notably for the gilt-head bream - Pagrus major). These trials started in the 70s, but have not proved very conclusive. The fish get used to the signal and gradually move away from its source. Though the idea is intellectually attractive, the method involves several unknowns which need to be clarified for each species and on each site:

- Conditions and cost of acquisition and conservation of the behaviour (control of habituation); if the natural food has to be permanently complemented then ranching looses its interest;
- Risk and cost of conditioning undesirable species (predators and commensals);
- Coherence between the spatial scale of the project, the economic and social structure of the operation (public, collective or individual), and the ownership regime of the stock and of the colonised ecosystem;
- Distribution of the costs and benefits between the protagonists of the project, the exploiters of the stock, and the other users of the ecosystem and of the space.

In the present state of knowledge, the method appears to be essentially a refinement of already mastered systems. Focusing on technical innovation, it under-estimates the institutional reforms on which its success depends.

Stocks are also confined, or it is planned to do so, in large natural spaces (lagoons, bays) by physical means (dams, nets, electric screens, bubble screens, etc.). This method also underlies the valliculture traditionally practised in Mediterranean lagoons. It is very developed in Asia, traditionally in fresh waters, more recently in coastal zones. As compared with physical barriers, bubble screens or electric screens has the advantage of not being subject to fouling or the accumulation of debris (algae), nor to damage by bad weather. On the other hand, their effectiveness remains to be proven. Some people consider this to be the way forward for aquaculture, in so far as the method would permit the extension towards the sea of intensive systems whose development in the coastal zone is blocked by constraints of space and conflicts of use, or is likely to be in the near future.

Conditions Required for the Take-off of New Extensive Systems

Production of Quality Juveniles

Although the small size of the eggs of the great majority of aquatic species complicates artificial reproduction, hatchery techniques have improved considerably over the past thirty years. In aquacultural research, the mastery of this function has often been considered a priority, even if artificial reproduction is not always essential for startup extensive cultivation (cf. shellfish culture or the transplantation of young plaice). Today the range of species whose reproduction is well or fairly well mastered is relatively large and given present knowledge and know-how this list could be fairly rapidly extended in the future. The production of juveniles is thus not a major obstacle to the development of extensive aquaculture.

On the other hand, the interest of artificial reproduction (uniformity of the product, reduction of natural hazards, less dependence on seasons and genetic selection) grows with the development of extensive systems. Even if the potential applications are less diversified than in intensive cultivation, genetic selection may contribute to the fight against epizootic diseases, improve growth performance, or serve to mark the individuals released (to help evaluate survival rates or genetic mixing with wild stocks). But selection also creates risks for the integrity of the genetic heritage of natural populations. At present, the lack of basic knowledge about the genetics of certain taxonomic groups, such as molluscs, is a constraint for selection.

Forcing Recruitment

For most of the species which have been the subject of prolonged experimentation there is an abundant literature describing:

- The qualities that the released individuals must possess: origin of the stock (natural or hatchery juveniles, genetic heritage as a function of the ecosystem to be colonised), size and age, adaptation to the life in the natural habitat (in the case of salmon, for example, prior acquisition of the reflexes and behaviours of feeding in the wild and defence against predators);
- The conditions for the release of individuals: habitat, date with respect to the smolt stage for salmon, seasonal trophic enrichment of the milieu, occurrence of predators, etc., with, as a possible corollary, operations to prepare the site (elimination of predators such as starfish in scallop cultivation in Japan, for example).

But, even for a single species, the conclusions are not always very concordant. This is explained by the essentially zootechnical approach often preferred in R&D programmes. In the absence of appropriate knowledge on the ecology of juveniles and the reproduction strategies of populations in their environment, it is difficult to define the qualities that the alevins have to exhibit and the conditions necessary for their release.

The constraints of aquatic life - and in particular dispersion - impose complex demographic strategies on marine populations. These strategies exhibit a number of common traits however:

- Very high fecundity, to compensate for the enormous mortality suffered by the eggs and larvae whose means of locomotion and food reserves are limited;
- The dependence of each population on a succession of hydrodynamic structures and the selection of physiological processes and behaviours adapted to the conditions obtaining in the structures occupied in the course of the successive stages of the life cycle (eggs, larvae, juveniles and adults);
- Homing behaviour: to complete their life cycle and ensure the survival of the population, and not simply of individuals, a sufficient number of reproducers have to return periodically to predetermined places in their habitat. The return to fry ponds which are stable in time and space thus appears to be a behavioural response to the double imperative of sexual reproduction and life in the aquatic milieu.

These constraints mean that marine species are made up of populations which are distinguished by the succession of patterns of distribution and migration specific to the different stages of the life cycle, their demographic characteristics (average abundance, growth, natural mortality), and their genetic heritage. During the initial stages, the numbers in each population fall dramatically under the effect of environmental processes independent of the abundance of the population (effect of temperature on sexual maturation, gametogenesis and viability of the eggs as in the case of the scallop; drifting into inhospitable hydrodynamic structures where the young stages may either perish or survive but lose any chance of participating in the reproduction of the population, as in the case of several species of fish), and trophic processes depending on the density of the population (predation, famine).

These demographic strategies have very important consequences for the design of high quality research programmes:

- As the regulation processes are concentrated in gametogenesis and the initial stages of the life cycle and are, at least partially, independent of density, it is very probable that for a good many species the biotic capacity of the habitat occupied by the adults is not a limiting factor for the levels of recruitment commonly observed for natural populations; for this reason substantial overpopulation should be compatible with the trophic capacity of the ecosystem. This hypothesis is confirmed by experience (shellfish culture, salmon ranching), as well as by the great amplitude of natural recruitment in certain species;
- From the strictly ecological standpoint, the general nature of homing behaviour and the fact that a majority of marine species reproduce in coastal areas allows us to conclude that sea ranching should be possible for a great variety of species;

- The technical failure of many experimental programmes is probably explained by inadequate taking into account of the particularities of the reproduction strategies of marine populations; massive mortality occurs when the individuals released do not find shelter (lobster, abalone), or the food they need, or fall prey to natural predators (scallop in France, lobster and cod in Norway, etc.);
- The choice of the stock is often critical for the success of restocking and acclimation programmes: as a general rule, the success rate for salmon ranching operations falls when the restocking is made with stocks coming from another ocean or the opposite side of the same ocean. The stocks have to be adapted to the length of the river or stream in order to be able swim up it. Certain poor rates of return are explained by a poor acquisition by cultivated animals of homing behaviour, a component of which is probably genetic. The reproduction strategy of the scallop, for which the success of recruitment depends on synchronisation between gametogenesis and the summer warming cycle in the habitat occupied by each population, could well cause restocking with spat from populations living in very different habitats to be a failure;
- On the other hand, as both juveniles and adults of diverse species may survive in hydrodynamic structures different from those inhabited by the populations to which they belong, and are more tolerant than eggs and larvae, the ecological constraints should generally be less severe for fattening and overpopulation operations not involving the participation of the released individuals in reproduction. Shellfish cultivation in sectors very distant from their area of reproduction is consistent with this hypothesis.

As the ecological constraints change with the objectives, the success of the R&D programmes thus depends on the explicit definition of the objectives, the underlying hypotheses and the experimental protocols. These conditions are difficult to meet in projects where research and development are not clearly distinguished. With advances in knowledge about the reproduction strategies of marine populations and the interest shown by marine ecologists in the full scale ecological experiments constituted by programmes to develop extensive systems, as well as in the possibilities for experimentation on the ecology and physiology of reproduction offered by hatcheries, these aspects are now better taken into account in extensive aquaculture programmes.

Availability of Sites

Competition for sites with already established uses of the coastal strip may be an obstacle to the implantation of new fish farms. Shellfish culture initially developed on the intertidal zone, in bays and basins whose shores were not particularly sought after at the time. In countries where shellfish culture has been implanted for a long time (Spain, France, Netherlands, etc.), a good proportion of the favourable sites have now been colonised. A limited extension of cultivation towards the open sea remains possible thanks to technical innovations (oysters cultivation in deep water, mussel culture on nets). Where it is well established, shellfish culture benefits from the advantage of its anteriority in the allocation of new concessions. This is not the case with new implantations (mussel culture on hurdles) which often comes up against the opposition of local residents. Environmental protection arguments may be put forward with greater force against new forms of cultivation than against traditional uses (fishing, agriculture). These practices reveal the weakness of current methods of allocating the maritime space between competing uses. The fact that several types of extensive cultivation (vagile species and deep sedentary species) make relatively little demand on the maritime space could be an advantage in regions where the coastal strip is very sought after.

Biotic Capacity of the Ecosystems and Water Quality

While, for the reasons explained above, the biotic capacity of ecosystem is generally not an obstacle for the start-up of new extensive systems, this observation is not an absolute rule. The productivity of the ecosystem is not sufficient everywhere to permit growth rates compatible with profitable shellfish farming. Urban and agricultural wastes may, within certain limits, remedy this poverty, just as certain clean-up programmes may reduce the productivity of certain shellfish farming basins.

More critical for the development of extensive aquaculture, and more particularly of shellfish culture, is the quality of the water. Feeding through filtration and bioaccumulation in the food chain make shellfish farming stocks and products very vulnerable to any deterioration in the quality of coastal waters under the effect of urban, agricultural and industrial pollution.

Technical Efficiency and Profitability

To evaluate the technical efficiency of a new system, it is necessary to be able to identify the individuals when they are recaptured. Several marking techniques are used for this purpose: external and internal physical marks, genetic marking (Norwegian PUSH programme for cod), or morphological marking (in Japan, for example, hatchery products may have characteristic traits: special form of the nostrils of sea bream, black pigmentation of the blind face of flounder, etc.). The possibility of changes in the rules of access to the stock and the imposing of a contribution to the financing of alevin rearing may cause fishermen to under-declare the recaptures. With good statistics, the estimation of the impact of rearing programmes on the biomass of a population and the fishing yield does not present any technical difficulties. On the other hand, the segregation of the different causes of mortality and losses (natural mortality, excess mortality of the artificial stock, straying rate, captures by different fleets) may prove difficult.

The existence of a persistent shortage of sea products of high commercial value in the rich countries and the fact that the development of extensive methods is not generally any more difficult for noble species than for others are factors favourable for the development of extensive aquaculture. On the other hand the poor correspondence, frequent in extensive systems, between social and private costs and benefits is a major handicap. It is the primary cause of the apparently poor efficiency of many extensive aquaculture programmes, and of the frequently insuperable difficulties encountered in trying to move from public to private financing. Demonstration of the technical and economic viability of a new system rarely suffices to ensure its take-off.

In order for a programme to be profitable, it suffices in principle for the price of the fish recaptured to exceed the cost of production and recapture of the individuals released. The shortcomings of the exclusion regimes mean that catches by third parties have to be counted as losses (in addition to those due to natural mortality and the straying of individuals), whereas from the technical standpoint they correspond to gains. A not insignificant proportion of both the gains (benefits for the conservation of piscicultural fauna) and the costs (use of the natural biotic capacity, risk of genetic mixing, modification of the landscape, etc.) are social. The financing will be all the more difficult to transfer to the direct private beneficiaries the greater these external effects. This is why a simple financial analysis at the level of the enterprise is not enough to establish the economic and ecological value of a new system. A cost/benefit analysis on the scale of the programme as a whole is essential. It will gain from being complemented by a risk analysis at the level of the main actors concerned (direct beneficiaries and other users of the ecosystem).

Ownership of the Stocks

Examination of extensive systems reveals the existence of correlations between the economic and social organisation of the operation and the process of intensification similar to those described for agriculture. Virtually all restocking programmes for the conservation of the fauna are financed by the public sector and benefit the society as a whole without restriction. The same is generally true of breeding young fish to support commercial fisheries in which access remains largely open and free: it is almost always financed by public funds. A partial move towards privatisation is to be seen in sport fishing in which anglers enjoy exclusive rights – collective or individual -, in return for a contribution to the financing of the breeding programmes. Elements of two systems are found in traditional fisheries: even though fishermen may contribute, the cost of the restocking programmes is generally borne by the state, but in common law access is often limited by a system of collective territorial rights. The collective ownership regime existing in Japan provides a good example of this type of regime. Japan is one of the rare countries to have legalised the privileges that coastal fishermen derive from their anteriority of occupation of the coastal strip and exploitation of coastal stocks. Collectively, fishermen's cooperatives and certain private groupings hold rights of quasi-ownership over halieutic resources. While they are not permitted to sell their rights, neither may the state withdraw these rights. This formalisation of the common law is not unconnected with the development of extensive aquaculture in this country. While private ownership and exploitation are the exception in the extensive cultivation of vagile species (private farms for exploitation for profit by the owner, or sport fisheries where access is governed by individual rights), they become the rule in shellfish culture, in which the stock is readily appropriable and effectively appropriated. The difference does not end there: shellfish culture is considerably more developed than the ranching of vagile species. While the statistics on this latter are disparate, the figures available show that national productions are often an order of magnitude lower than those of shellfish culture. Lastly, the private ownership of the stock is virtually the rule in intensive fish farms.

Counter-examples of programmes which, despite their technical success, have not led to commercial production, or whose the take-off is blocked by institutional constraints, are no less clear. In France, for example, where the extensive culture of the scallop is technically mastered and has been the subject of full-scale experiments financed by the public sector, the transfer of the cost of planting to private operators is progressing only slowly. The trade organisations accept the fact that fishermen should make a contribution, on a voluntary lump-sum basis, to the financing of a collective restocking programme, but they are opposed to the granting of collective concessions on natural beds – though these are on the face of it the most promising -, or to modulation of the individual contributions which would give fishermen rights of recapture proportional to their contributions. This would in fact result in the adoption of a system of tradable individual quotas to regulate access to the fishing of natural beds. The start-up of cultivation on individual concessions comes up against the scale of lump-sum charges applied by the administration for the cultivation of oysters and mussels. Would-be fish farmers consider that the scale in force is too high for starting up a new system of cultivation where the level of risk is still high.

Thus the privatisation of the stock appears to be a determining factor for the process of intensification: from fishing to intensive cultivation, individualisation of exploitation (traditional groups, sport fishing associations, commercial fleet owners, private farms), technical intensification (conservation, fishing, extensive aquaculture, intensive aquaculture), privatisation of the stock (public, collective, and individual ownership), and passage to the monetary sphere (conservation, sport fishing, commercial fishing, aquaculture) all go hand-in-hand.

In primary production where the allocation of the factors is governed by elaborate systems of private ownership, the rights explicitly define the variables, the holders, the quantities, the offences and the sanctions. They are exclusive in the sense that all the gains and losses inherent in the exercise of a right fall to the holders of the right. They are combinable and for this reason tradable. They are technically applicable and effectively applied, since a right which is not applied is not a right.

The mobility of vagile species makes it impossible to simply transpose to extensive cultivation the solution adopted in agriculture and shellfish culture – i.e. the subdivision of the area - for the appropriation of the stock. However, a better coincidence between the distribution of costs and benefits may be sought by allocating exclusive quantitative rights of alevin rearing and recapture to producers. However the application of this system in cultivation where the same species is fished in the same areas comes up against a problem: the impossibility of distinguishing the cultivated or wild origin of the captures. This uncertainty implies that fishing and cultivation should be subject to the same system of regulation. But it is by no means certain that the increase in production expected from a new system of cultivation will suffice to convince fishermen, who have open and free access to the wild stocks, that it is in their own interest to adopt a system of quantitative individual rights.

The adoption of individual rights also raises the question of the choice of the allocation mechanism. With the opening up of rural societies to national economies, the systems of customary rights based on social control are gradually falling into disuse. The poor performance of traditional methods of regulation of access in commercial fishing illustrates the ineffectiveness of administrative regulation as a way of allocating a factor of production in a commercial activity. The advantages of tradable rights (determination of the price of fishing rights by direct negotiation between the parties) comes up against the opposition aroused by the adoption of this system for allocating factors hitherto allocated by non-market mechanisms, and objections of inequality resulting from the people's different capacities for taking advantage of the new opportunities created by the adoption of a new system of regulation.

The evolution of ownership regimes and mechanisms for trading rights finally depends on the clarification of national jurisdictions which, at sea as on land, define and guarantee the ownership regimes for natural resources. While the new law of the sea does indeed grant states' rights of quasi-ownership over the resources contained within their exclusive economic zones (EEZs), their authority remains diluted for the stocks whose area of distribution covers the EEZ and the high seas, and is shared for stocks whose area extends over several EEZs.

But for a few exceptions - that of Iceland in particular, which is distinguished by its geographical isolation and the initiatives that this country has taken to adjust the organisation of its fisheries to new conditions -, the present situation in Europe scarcely seems favourable for the extension of extensive aquaculture to vagile species. The great development of continental waters for the moment reduces the prospects for highly migratory species like salmon. In the majority of

European fisheries, access to the resources still remains for the most part open and free. In the EU, the Common Fisheries Policy has so far paid more attention to the maintenance of the status quo and the minimisation of the immediate consequences for fishermen of the recurrent measures taken for the conservation of stocks, than to the clarification of the regime of ownership of the resources and the modernisation of the mechanisms for regulating access.

Rationalisation of Mature Systems

The constraints that extensive systems come up against when they arrive at maturity differ in nature from those which impede their take-off. Because shellfish culture is one of the small number of well-developed systems, our examination of the opportunities and conditions for the rationalisation of mature systems will be based on this mode of production. But the conclusions are transposable to the other systems.

Adjustment of the Stocks to the Biotic Capacity

The hypothesis according to which the trophic capacity of the marine ecosystem should be able to support biomasses greater than those of wild populations does not permit us to predict an increase in the biomass supportable by the various aquacultural ecosystems. The recruitment level of a population and the biotic capacity of an ecosystem to support the juvenile and adult fractions of a population in fact concern different stages of the life cycle and result from distinct processes. Thus the example of shellfish culture shows that the aquacultural potential varies considerably from one coastal ecosystem to another. But whatever the trophic capacity of an ecosystem may be, it is always limited. The finite nature of the natural resource will become increasingly apparent as the biomass in cultivation increases, and will result in a progressive decline in yield growth.

It is thus that in semi-enclosed shellfish farming basins where the water is renewed only slowly, the primary productivity (microphytobenthos and phytoplankton) that the stock feeds on becomes an increasingly limiting factor as the biomass of the stock grows. For access to food for their stocks shellfish farming enterprises find themselves in a situation of competition analogous to that well known in fishing. The continuous movement of the water within a basin prevents the shellfish growers from separately adjusting their stock to the productivity of their concessions. To increase the share of the primary production of the basin that they appropriate they individually have an interest in increasing their stocks, despite the overloading of the basin and the overall decline in yields that this behaviour leads to. Thus, in the Marennes-Oléron basin, the maximum production – slightly over 40,000 tonnes - is obtained for a biomass in the order of 80,000 tonnes. With the Portuguese oyster, the total biomass reached 200 000 tonnes. Exactly the same process was repeated when the basin was replanted with of the Japanese oyster following the disappearance of the Portuguese oyster. The overstocking slows the growth of the oysters - which has the effect of lengthening the fattening cycle (from two to five years) -, and increasing the general mortality rate (multiplying it by two or three). The resulting loss of income only reinforces the overstocking behaviour on the part of oyster growers.

Preservation of Ecosystems

In sectors where shellfish culture is very developed, the farms may modify the circulation of the water (hurdles) and the substrate (sedimentation by the biodeposits - faeces and pseudofaeces).

The productivity of the sites may be affected. In the Bay of Mont St Michel, for example, the deterioration in performance has led the growers to adopt self-imposed limitation of the stocks (spacing of the hurdles).

Epizootic Diseases

The recent history of French shellfish culture illustrates the risk of epizootic diseases connected with the transfer of spat and shellfish between countries and between basins. Between 1967 and 1972, two viruses totally eradicated the Portuguese oyster stocks. Fraudulently introduced, the Japanese oyster is thought to have been the resistant vector of the infectious agent. Through reducing the resistance of the stocks, the overstocking of the basins was probably a facilitating factor, but not the cause of the epizootic diseases. The epidemics appeared, in fact, after peak production levels had been reached in the main basins.

The propagation of diseases has been facilitated by transfers of shellfish between basins. In 1978 and 1982, the flat oyster stocks were hit in their turn by two epidemics whose economic and social consequences were catastrophic. Production fell from over 20,000 tonnes to just a few thousand tonnes. Enterprises had to change species and farming practices and were forced to concentrate. Marketing circuits were disrupted and changed to the benefit of the traditional producers of cupped oysters. For the Brittany region alone, the loss of turnover for the period 1974-82 has been estimated at 1.6 billion current francs.

While the replacement of the Portuguese oyster by the Japanese has made it possible for production to start anew, the present monoculture increases the risk of epizootic diseases, which is all the more serious in that at present there is no substitute species. The difficulty of treating stocks in farms where curative measures are possible only on the spat before its immersion or, with heavy handling costs, on adults in the fattening basins, and the weak advance of mollusc genetics, mean that health strategies depend essentially on prevention. Checks are carried out on imports and also on immersion, transfer between basins, and farming practices and the state of health of the stocks are monitored.

Exotic Species

Despite these controls, exotic species are regularly introduced on the occasion of imports of spat and illegal immersion of shellfish before their commercialisation. The apparitions of non-indigenous are in fact concentrated around the big oyster growing basins: Hydroides ezoensis, Apitasia pulchella, Anomia chinensis, Balanus amphytrite, etc. on the Atlantic coast of France; algae Laminaria Japanica, Undaria pinnatifida and Sargassum muticum in the Thau coastal pond on the Mediterranean coast of France. These imports may also have helped propagate the shelf limpet (Crepidula fornicata) - a competitor of the shellfish stocks - and a type of periwinkle (Ocenebra erinacea) - a predator of the oyster. Similarly, transfers of fish between Sweden and Norway were at the origin of the introduction of a monogene, Gyrodactylus salaris, which has provoked the catastrophic decline of certain wild salmon populations.

As shown by examples of the fortuitous introduction of exotic species, the risk for biodiversity is particularly serious when the plantings concern non-indigenous species. There have been many transplantation experiments in the past: Pacific salmon on the Atlantic coasts of Canada and the USA, Russia, Chile, New Zealand, etc.; Atlantic salmon on the Pacific coast of Canada; Atlantic lobster in the Pacific; king crab from the Sea of Okhotsk in the Barents Sea; South African crawfish in the Channel; algae (Laminaria, Undaria, etc.) from Japan or Macrocystis from the Pacific coast of America in the Channel, etc. The Japanese oyster (Crassostrea virginica) has colonised practically the entire northern hemisphere. To limit this risk, the International Council for the Exploration of the Sea and the European Consultative Committee for Inland Water Fisheries have published a code and a practical guide applicable to the introduction of non-indigenous species and to the transfer of stocks within the same country. These provisions are not binding however. The example of shellfish culture shows how difficult it is to ensure compliance with the regulations governing transfers once cultivation has reached a high level of development.

Genetic Mixing

Large-scale rearing and restocking programmes which are not rigorously planned may constitute a threat to biodiversity. Through diluting the demographic structures of a species, the invasion of an ecosystem by a common genome may reduce the genetic adaptability of the wild population and cause the disappearance of local populations. The risk depends on the survival and reproduction rates of the foreign individuals, as well as the hybridisation rate and competition of the exotic strains with the wild stocks.

The risk is higher with intensive cultivation, since selection is aimed at artificialising the genetic heritage of the stocks. In the salmon farms of northern Europe, escapes from fish farms in cages already commonly exceed the recruitment of the wild populations, and are capable of exceeding that of the sea ranching programmes: in Scotland and Norway, the escapes from cages are of the same order (2 million individuals) as the number of smolts deliberately released for sea ranching.

In the case of restocking with wild strains already present in the ecosystem colonised, the risk of genetic mixing is theoretically zero. To conserve the biodiversity of the wild populations of anadromous species such as the salmonids, planting may also be limited to one part only of the rivers, the wild populations being conserved in the others. This is what is done in Ireland for trout and salmon. Restocking may also be practised with sterile animals (triploids), but the genetic manipulation of strains intended for human consumption does not have a favourable image in the eyes of the public.

Up to now there is little evidence of mixing between domestic stocks and wild stocks and the modifications observed are mostly minor. In a review of the knowledge available on the genetic impacts of hatchery fish on the wild populations of salmon and steelhead trout in the Pacific, Campton concluded in these terms: "Although widely accepted theoretical and conceptual arguments suggest that the direct genetic effects of hatchery fish on wild populations could be substantial and potentially detrimental, the empirical data supporting those arguments are absent or largely circumstantial." The problem is that for a long time sea ranching programmes were conducted without regard to the identity of the populations, without first establishing the zero state of the genetic heritage of the wild populations, and without genetically marking the individuals released. Since then evaluations of the rate of mixing have been carried out with genetically marked populations, in particular on cod in Norway and trout in Ireland.

In the present state of knowledge, the risk is thus poorly understood. But the fact that public opinion is probably very sensitive to it is in itself a threat to the development of aquaculture. For this reason certain specialists put this risk at the top of the list of the impacts of intensive cultivation on the environment, ahead of organic wastes - faeces and unconsumed food -, antibiotics, medicines and antifouling paints used on the cages. In order to have precise instruments to measure and monitor genetic impacts on natural populations, the Convention on Biodiversity is now preparing an in-depth analysis of the risk created by sea ranching and intensive aquaculture.

Ownership of Natural Resources

When an extensive cultivation system becomes mature, its rationalisation involves the adjustment of the biomass in cultivation to the trophic capacity of the ecosystem. This adjustment has to be made on the scale of the entire ecosystem. This requires that the ownership of the ecosystem - the natural resource - should be clarified. To take an image from agriculture, the breeder should have exclusive rights over his livestock and its pastures in order to manage it properly. The transposition of this model to the sea comes up against to two difficulties, one technical, the other institutional:

- The disparity of scales between the unitary resource - the aquacultural ecosystem -, on the one hand, and the fish farms on the other: the problem may be overcome by attributing the function of adjustment of the total biomass to a single authority - public or collective; for this, the functions which are concerned with the ownership of the natural resource ought to be distinguished from those which concern the use of this resource - i.e. the exploitation of defined fractions of the natural resource;

- The disparity of the mechanisms used to allocate the human inputs (capital and labour), on the one hand, and the fractions of the natural resource, on the other. The former are mobilised by the fish farmers. For this they rely on mechanisms mainly of a market nature. Rights of use, for their part, are traditionally allocated by decisions of the authority responsible for development, in more or less close collaboration with the body of fish farmers. Being a matter of allocating a factor of production in a competitive activity, experience shows that these methods, whether they are used separately or jointly, are less efficient for controlling overexploitation than economic mechanisms - taxation or market mechanisms.

The problem of the clarification of exclusivity rights thus arises at two levels: that of the ownership of the natural resource, and that of the right of use. The problem can be analysed in same way for the other uses of the natural resource: fishing and intensive aquaculture. Only the fraction of the ecosystem used changes with different uses.

The need for clarification of the ownership regime is independent of the identity of the holders of the rights, as of the mechanism for allocating the rights of use. If the regulation of uses is in the hands of a public administration, its efficiency will be reduced if the respective responsibilities of public sector and trade structures, of sectoral administrations (fishing, aquaculture, environment), and different levels of the politico-administrative structure (from local to international), are imprecise. Seen from this angle, the concept of ownership no longer appears as an absolute right over things, but as an assortment of exclusive rights formalised to exercise different controls, allocated according to the respective capacities of the holders (from individual to international) to exercise controls pertinent to the appropriate scales.

While the adoption of the EEZ constitutes a decisive stage in the adjustment of the institutions to new conditions, the new law of the sea remains insufficient to ensure the conservation of renewable resources, the development of new aquacultural systems, and the organisation of fishing. The major shortcomings concern:

- Clarification of the exclusivity regimes:
  - National legislations still fail to make a proper distinction between the ownership and the right of use of natural resources;
  - This imprecision has an impact on the exercise of the functions which attach to these responsibilities:
    - ◊ The integrated development of resources is difficult to achieve when it is approached from the standpoint of sectors of activity rather than by complexes of resources: the take-off of the ranching of vagile species requires, as we have seen, the integration of the systems for the allocation of the rights of recapture of the cultured stocks and of fishing the wild stocks. Similarly, to be efficient the adjustment of the aquacultural stocks to the capacity of the ecosystem and the conservation of this capacity should be in the hands of the same structures; it is a matter, in fact, of allocating the potential of the natural ecosystem between competing uses;
    - ◊ The supervisory administrations have difficulty in reconciling their responsibility for the conservation of natural resources with that of supporting the professions. In the majority of European fisheries policies, for example, the minimisation of the immediate costs for the profession of the recurrent resource management measures often takes precedence over the introduction of the reforms necessary for the rationalisation of the sector and the conservation of halieutic stocks;
  - In many countries, the organisation of fisheries remains centralised at national level – in the case of the European Union it is at Community level, which is larger than the scale of the great majority of halieutic stocks and ecosystems. This centralisation reduces the efficiency of public interventions (subsidiarity principle). It was justified when the prime responsibility of the administrations was to defend the interests of their national fleets, free to operate in the world ocean, but this is no longer the case since the adjustment of the uses to the potentials of the natural resources has become the main consideration, and now that the extension of national jurisdictions has given national authorities the power to exercise their responsibilities on the scale of the unitary resources;
- Modernisation of the mechanisms for allocating rights of use: the origin of the dynamic of overexploitation being economic, economic mechanisms are generally more efficient for rationalising the commercial uses and, as a result, conserving the resources. For the same reasons, the development of aquaculture will be more dynamic if the rights of use of the natural resources are allocated by economic mechanisms.

Exogenous Factors

Deterioration of the Environment

Shellfish culture provides a good example of the conflicts which arise between extensive aquacultural systems and the utilisation of the carrying capacities of the milieu. These conflicts are not symmetrical. While shellfish culture modifies the environment very little, its localisation in coastal zones where the wastes of human activities accumulate, and the fact that the cultivated species feed by filtration, which concentrates any micro-organisms or toxic substances in the products, makes them very sensitive to any deterioration in the cultivation milieu. This vulnerability may be illustrated by the following examples.

The growth of the stocks and their state of health depend on the quality of the shellfish culture ecosystem:

- At the end of the 70s, the use of tributyl tin paints to protect the hulls of ships and boats against biological fouling caused oyster production in the Arcachon basin (France) to fall from 12,000 to 3,000 tonnes; production returned to its previous level after the use of these paints was banned;
- Plankton bloom (Gyrodinium aureolum) periodically causes high mortality in mollusc cultivation (scallop hatcheries and stocks, for example).

Eating shellfish raw increases the health risk. Purification techniques are, in fact, efficient or sure only for microbial contamination, and not for chemical or biotoxin contamination. This health risk is also liable to lastingly affect the quality image of the products:

- Epidemics caused by the consumption of shellfish may be of bacterial origin (cholera occasioned by the consumption of mussels as in Naples in 1973; typhoid fever; dysentery, etc.) or of viral origin;
- Plankton bloom may synthesise neurotoxins capable of causing paralyses that can be serious for man (Protogonyaulax for example) and toxins capable of causing less serious troubles such as diarrhoea (Dynophysis).

What determines the appearance of toxic bloom has not yet been established. These blooms appear naturally, in fact, but the apparent increase in their manifestations and the concentration of occurrences in the plumes of rivers lead to the suspicion that the origin must be partly anthropic.

Demand

There is a considerable deficit of sea products in the majority of European countries, especially quality products. Given the high nutritive value (low calories, low fat and cholesterol contents) and taste quality of fish, the demand for these products in unlikely to fall. The deficit could even increase if European and world halieutic production continues to decline and if the supply from developing countries levels off. Given the predominance of fishing in the total supply and the time necessary for the institutional adjustments required in the organisation of fisheries, this scenario is very plausible. Even if the expansion of existing fish farms and the take-off of new intensive projects reduce the deficit, diversification will probably not go hand in hand with the increase in production, because intensification is generally accompanied by a reduction in the diversity of the species cultivated. The potential interest of extensive aquaculture should be viewed in this light. Technically, the method can be used to produce a great variety of species (shellfish, prawns, turbot, bream, sturgeon, salmonids, etc.), and the difficulties of development are on average no greater for systems producing species of high commercial value than for any other species.

Extensive aquaculture could also benefit from a better image than that of fishing and intensive aquaculture. As demonstrated by the actions of ecological movements against the fishing of culturally sensitive species, the inability to prevent overexploitation and to manage the conflicts within the fishing industry are starting to tarnish the excellent image this activity once enjoyed. The image of intensive aquaculture suffers from the damage it is accused of doing to the environment (deterioration of landscapes, milieus and aquatic populations). Its capacity to produce quality products is also called into question, an opinion that the recent mishaps in agriculture have done nothing to correct. The high consumption of fishmeal, fresh water (for the production of smolts in salmon culture) and energy in intensive aquaculture may also be detrimental to its image. By comparison, see ranching could appear to be a more environmentally-friendly activity, and one capable of producing a more natural type of food (provided that the quality of the environment is preserved at the same time). Another aspect that should not be ignored in countries where leisure activities are becoming increasingly important is that it may also contribute to the development and diversification of recreational activities.

The health risk and the difficulty of controlling the quality of the products have so far constituted a serious obstacle to the development of international trade in shellfish. The harmonisation of national quality control legislation being pursued by the European Commission should reduce this constraint.

There is thus reason to believe that extensive systems will enjoy strong demand and generally be looked more favourably than the other modes of production. But this conclusion does not mean that such systems will be able to live up to this expectation in practice.

Role of the Public Sector

Research

The development and installation of extensive systems gives rise to complex questions that research, if it is efficient, may help to resolve. The public sector has a particular role to play in the production of knowledge about issues whose scale takes them out of the field of interest of private operators:

- Restocking programmes for the conservation of piscicultural fauna and halieutic stocks: this is typically the case with support programmes for salmonid populations;
- Research and expertise to support the rationalisation of shellfish farming systems: conservation of the biotic capacity and of the environmental quality, adjustment of the biomasses in cultivation to the trophic capacity, fight against epizootic diseases, protection of consumers' health;
- Research on the prospects and conditions for the regulation of access to natural marine resources.

Our examination of the conditions required for take-off and the optimisation of extensive systems highlighted the variety of the disciplines potentially useful for the study of these questions:

- Ecology: determination of the recruitment and regulation strategies of natural populations, biotic capacity of aquacultural ecosystems, carrying capacity of aquacultural ecosystems, genetic mixing and contamination between aquacultural stocks and wild populations, fight against epizootic diseases, determination of the causes of plankton bloom, etc.;

- Zootechnics: artificial reproduction, pathology and prophylaxis, genetics, etc. of the species cultivated;
- Economics: cost/benefit analyses of extensive systems, economics of aquaculture and the environment;
- Law of the sea and of the environment: regulation of access to natural marine resources;
- Halieutics – in the agronomic sense of the term -: process of domestication in the marine domain.

It revealed above all that progress in extensive aquaculture depends on the capacity to provide overall responses to complex and coherent sets of questions which differ according to the nature and state of advance of the production systems considered. The lack of coverage and coherence of the research programmes may be considered to be the main cause of past failures. For a long time, targeted research was concerned above all with small-scale questions (such as artificial reproduction) and tended to pay less attention to larger questions (such as the regulation of the natural populations in the aquatic milieu in the biological sciences, or the process of technical intensification, the regulation of access to resources and the role of the ownership in the development and use of renewable marine resources in the human sciences).

Significant efficiency gains may thus be expected from adopting a multidisciplinary approach including the following elements:

- Explicit definition of the objectives: restocking, acclimation, fattening with overpopulation;
- Analysis of culture systems to identify and rank the conditions required for achieving the objectives of development and organisation;
- Identification, conceptualisation and articulation of the pertinent research topics;
- Formulation of hypotheses and of experimental protocols for testing them;
- Execution of the investigations;
- Analysis and integration of the results.

Two other causes of inefficiency frequently cited in the literature concern the lack of rigour in the experimental protocols and the confusion between research and development. These two shortcomings are not independent. While the association of development projects and experimental programmes makes it easier to obtain research funding, this type of funding often hampers the establishment of rigorous experimental protocols and leads to the abandonment of the investigations when the results do not confirm the initial hypotheses.

Research on extensive systems may contribute to the advancement of knowledge on questions of great scientific interest: regulation of marine populations, domestication and technical intensification of aquatic resources, the role of institutions in the exploitation and conservation of these resources. Because they permit full-scale experimentation on the ecology of marine populations, major extensive aquaculture programmes, like those implemented in France on scallops or in Norway on cod, are likely to be of interest to basic research. This interest may facilitate the mobilisation of theoretical and methodological competencies necessary for the study of new questions concerning the exploitation of the living resources of the sea.

Control of the Quality of the Environment and of Aquacultural Stocks and Products

The scale of the controls necessary and concern for the protection of consumers' health justify the intervention of the public sector in three other fields: conservation of environmental quality, preservation of the health of the stocks, and quality control of the products. These three fields are connected with one another. The fact that the state of health of the stocks and the wholesomeness of the products depend upon the quality of the environment explains why the European countries concerned themselves very early – right from the beginning of the century in some cases – with setting up monitoring systems for milieus, stocks and products of cultivation. In 1991, a European directive harmonised the legislation of the EU member states in this field. A system of four levels of wholesomeness was instituted. It has three objectives: conservation of the cultivation milieu, prevention of epizootic diseases, and consumer protection.

Table: New classification of oyster growing areas applied since 1 January 1996 in the European Union.

Areas	Upper limit	Exploitation	
		Natural beds	Cultures
Fit for consumption	300 faecal coliforms/100 g of flesh	authorised (direct consumption)	authorised (direct consumption)
Exploitable	6 000 faecal coliforms/100 g of flesh	authorised (relaying or purification)	authorised (relaying or purification)
Exploitable	60 000 faecal coliforms/ 100 g of flesh + heavy metals + pesticides	authorised (relaying - > 2 months - or purification)	prohibited (except in the case of derogation)
Not fit for human consumption		prohibited	prohibited

Being reactive, these measures do not address the causes of the degradation of the quality of the coastal milieu or inland seas, only the effects. The passage to preventive requires the adjustment of the institutions.

Adjustment of the Institutions to New Conditions

In the maritime field, technical intensification and the rationalisation of production are hampered by the regimes regulating access to the natural resources. Recent historical studies in the fields of economics and sociology stress role of institutions, and in particular of ownership regimes, in economic growth. For some commentators they are even considered to be fundamental: ' ... innovation, economies of scale, education, capital accumulation, etc..., are not causes of growth, ... Growth will simply not occur unless the existing economic organisation is efficient. Individuals must be lured by incentives to undertake the socially desirable activities. Some mechanisms must be devised to bring social and private rates of return into closer parity ... A discrepancy between private and social benefits or costs means that some third party or parties, without their consent, will receive some of the benefits or incur some of the costs. Such a difference occurs whenever property rights are poorly defined, or are not enforced' . This theory is particularly relevant in the maritime world where, as compared with terrestrial regimes, the institutions governing access to natural resources still have difficulty, despite recent advances in the law of the sea, in adapting to the new situation of the scarcity of natural resources.

At both national and international level, collective initiatives to regulate access are bound to remain uncertain so long as they are not formalised. The problem is that collective decisions do not commit third parties. This is why, over the centuries, when it has been a matter of assuring regalian functions (army, justice, police and money) on which the regulation of access depends, public solutions have always ended up by taking precedence over individual or collective initiatives. To limit theft and freeloading, the application of a common regime has always turned out less costly in the end and public solutions have prevailed.

But the expectation of considerable social benefits does not mean that the institutions necessary for their concretisation emerge spontaneously. Institutional innovations differ in this respect from technical innovations. While the latter may cost a lot, their exploitation is often within the reach of private initiatives, and their benefits go to individuals who are prepared to take the risk. Institutional reforms, for their part, come up against the opposition of vested interests, the inertia of the bureaucracies, and people's apprehensions about the social consequences of change. These constraints have a high immediate cost. But above all, the potential benefits fall to society as a whole, and not to the politicians or administrations who would be prepared to promote them. Under these circumstances advocates of reform are thin on the ground. While the solutions adopted by societies for their organisation may be rational, they do not necessarily seize the opportunities presented to them. The great geographical and sectoral disparities in economic development are indicative of the non deterministic nature of the evolution of societies. Because the adjustment of institutions is part of the political process and depends on public initiatives whose realisation is not a foregone conclusion, the public authorities should direct their efforts at the adjustment of the institutions to new conditions, rather than the support of production and the defence of professional interests.

Intensive Aquaculture

In a semi-intensive system, the production of the pond is increased beyond the level of extensive aquaculture by adding supplementary feed, usually in the form of dry pellets, to integrate the feed naturally available in the pond, allowing for higher stocking density and production per hectare.

In intensive farming, the fish are kept at too high a stocking density to obtain significant amounts of feed from their environment. Instead the fish are dependent on the feed provided and water must be replenished at a high rate to maintain oxygen levels and remove waste. The levels of feed inputs and management of the water affect the stocking density of the fish that can be supported.

In the EU, 80% of farmed fish production comprises the following four species that are predominantly intensively farmed: rainbow trout (reared both in freshwater and at sea) and marine fishes Atlantic salmon , gilthead seabeam and European seabass. In Norway, Atlantic salmon and rainbow trout represent 98% of farmed fish production. Marine fish and rainbow trout are dependent on fish meal and/or fish oil feed. Together they form a relatively small proportion of farmed fish production tonnage, but consume a large part of the global fish oil and fish meal consumed by aquaculture. The intensive farming of these fish, whereby large numbers of them are confined in a small area, causes a range of serious welfare problems.

Pangasius catfish is farmed very intensively in Vietnam, where half of production is for export, the EU and US being leading importers. Vietnamese production of pangasius has increased 10-fold in the last decade to 1.1 million tonnes in 2010 (78% of global farmed pangasius production), of which 0.66 million tonnes was exported. Extremely high stocking densities are made possible by a high rate of water exchange and by the ability of pangasius to breathe atmospheric oxygen, which makes them able to tolerate low levels of dissolved oxygen and highly polluted water. Small-scale pond polyculture systems are being replaced by intensive monocultures that rear pangasius in ponds (which exchange water with nearby river tributaries by tidal exchange and pumping) and in net cages and net pens sited on major river tributaries of the Mekong River delta. Intensive monoculture ponds and net pens are usually stocked at 40-60 fish per square metre, ponds sometimes even higher. Stocking densities for net cages are typically 100-150 fish per square metre.

According to one study, in Vietnam, pangasius are grown in deep ponds (4 m^2 deep) to a market size of 1 kg at stocking densities of 44 fish per m^2. Death rates for these fish are reportedly 20-25%. According to the same study, in Bangladesh, stocking densities for this species are much lower at 6 fish per m2 in 1m deep ponds, and death rates are significantly lower at 10%. Whereas in Vietnam pangasius are farmed in monoculture, the vast majority of farms in Bangladesh also stock around 10–20% filter-feeding carps which eat up algal blooms produced in the fertile ponds (made so by the feed inputs). These carp require dissolved oxygen and are far less tolerant of poor water quality than pangasius, indicating that water quality is maintained at relatively high levels. According to the authors of this study, the better levels of fish welfare (with respect to stocking density, water quality and mortality rates) in Bangladesh also resulted in less use of antibiotics. Both farm-made and commercial feeds, containing wild fish, are widely used in Bangladesh and Vietnam.

Grass carp intensively reared in cages are stocked with other carp (usually as the major species) and the cages are usually about 60 m^2, with a depth of 2-2.5 m. The fish are fed with aquatic weeds/terrestrial grasses and pelleted or other commercial feeds. Grass carp are stocked at 10 to 20 fish per cubic metre. In addition Wuchang bream are also stocked at 35 to 50 fish per cubic metre. A small quantity of silver and bighead carp are also stocked (1% of the total fish) as 'cage cleaners' (to eat and control algae).

Intensive Sea Farming

Sea cages hold fish captive in a large pocket-shaped net anchored to the bottom and maintained on the surface by a rectangular or circular floating framework. They are widely used for rearing finfish, such as salmon, sea bass and sea bream, and to a lesser extent trout, in coastal and open waters, in areas sheltered from excessive wave action, with sufficiently deep water and relatively low current speeds. Several cages are typically grouped together in rafts, often housing moorings and walkways for boat access, feed storage and feeding equipment. As the water flows freely to the cages, the openness of the system makes it vulnerable to external influences (i.e. pollution events or physical impact) as well as exposing the adjacent environment to the stock, and the fish farm effluents.

Recirculation systems on land can also be used for the farming of marine species.

Fish Farms

A fish farm is where fish are raised for food. This area of agriculture is known as aquaculture. The fish are raised in tanks or in ponds and sold to people or companies.

Some fish that are raised on fish farms are:

- Tilapia,
- Cod,
- Catfish,
- Salmon,
- Carp.

There are different types of fish farms that utilize different aquiculture methods.

The first method is the cage system which use cages that are placed in lakes, ponds and oceans that contain the fish. This method is also widely referred to as off-shore cultivation. Fish are kept in the cage like structures and are "artificially fed" and harvested. The fish farming cage method has made numerous technological advances over the years, especially with reducing diseases and environmental concerns. However, the number one concern of the cage method is fish escaping and being loose among the wild fish population.

The second method is irrigation ditch or pond systems for raising fish. This basic requirement for this method is to have a ditch or a pond that holds water. This is a unique system because at a small level, fish are artificially fed and the waste produced from the fish is then used to fertilize farmers' fields. On a larger scale, mostly in ponds, the pond is self-sustaining as it grows plants and algae for fish food.

The third method of fish farming is called composite fish culture which is a type of fish farming that allows both local fish species and imported fish species to coexist in the same pond. The number of species depends, but it is sometimes upwards of six fish species in a single pond. The fish species are always carefully chosen to ensure that species can coexist and reduce competition for food.

The fourth method of fish farming is called integrated recycling systems which is considered the largest scale method of "pure" fish farming. This approach uses large plastic tanks that are placed inside a greenhouse. There are hydroponic beds that are placed near the plastic tanks. The water in the plastic tanks is circulated to the hydroponic beds, where the fish feed waste goes to provide nutrients to the plant crops that are grown in the hydroponic beds. The majority of types of plants that are grown in the hydroponic beds are herbs such as parsley and basil.

The last type of fish farming method is called classic fry farming this method is also known as "flow through system". This is when sport fish species are raised from eggs and are put in streams and released.

Cage System

The right choice of site contributes significantly in the success of cage farm. Site selection is vitally important since it can greatly influence economic viability by determining capital outlay, by affecting running costs, rate of production and mortality factors.

- Site selection is a key factor in any aquaculture operation, affecting both success and sustainability.
- Circular cages of different diameter ranging from 2 m to 15 m, designed for the culture of fishes such as milkfish, mullet, cobia, pompano, sea bass, pearl spot, shellfishes such as shrimps, crabs and lobsters were experimented and demonstrated successfully in India by Central Marine Fisheries Research Institute (CMFRI).
- Stocking of right sized fish juveniles in adequate stocking density is another factor which determines the success of farming. The stocking density and size of stocked fishes varies with different species.
- Proper feeding of quality feeds, periodic monitoring and cleaning of cages contributes immensely to the success of cage farming.
- With proper management of cage erected at an ideal location can yield a production of 20-40kg/m^3 with various species of fishes.

Cage aquaculture involves the growing of fishes in existing water resources while being enclosed in a net cage which allows free flow of water. It is an aquaculture production system made of a floating frame, net materials and mooring system (with rope, buoy, anchor etc.) with a round or square shape floating net to hold and culture large number of fishes and can be installed in reservoir, river, lake or sea. A catwalk and handrail is built around a battery of floating cages. There are 4 types of fish-rearing cages namely: i) Fixed cages, ii) Floating cages, iii) Submerged cages and iv) Submersible cages. Economically speaking, cage culture is a low impact farming practice with high returns and least carbon emission activity. Farming of fish in an existing water body removes one of the biggest constraints of fish farming on land, i.e., the need for a constant flow of clean, oxygenated water. Cage farms are positioned in such way to utilize natural currents, which provide the fish with oxygen and other appropriate natural conditions.

Site Selection

Different criteria must be addressed before site selection for cage culture.The physico-chemical parameters like temperature, salinity, oxygen, waveaction, pollution, algal blooms, water exchange, etc. that determine whether a species can thrive in an environment. Other criteria which must be considered for site selection are weather conditions, shelter, depth, substrate, etc. Finally legal aspects, access, proximity to hatcheries or fishing harbor, security, economic, social and market considerations etc. are to be taken care.

Cage Size

It is a fact that costs per unit volume decrease with increasing cage size, within the limits of the materials and construction methods used. CMFRI has developed open sea cages of 6 m dia and 15 m dia for grow out fish culture and 2 m dia HDPE cages for seed rearing. Ideal size for grow out cage is 6 m due to its easy maneuvering and reduced labour. For fingerling, 2m cages can be used.

Cage Frames and Nets

Different cage materials can be used for cage farmes. Materials commonly used are High Density Poly Ethylene (HDPE), Galvanised iron (GI) pipes, PVC pipes, etc. HDPE frames are expensive, but long lasting. Cost effective epoxy coated Galvanized Iron (GI) frames are recommended for Small groups and fishermen. GI frames have less life span when compared to HDPE frames.

Nets of varying dimensions and materials were tested for cage culture in India. CMFRI has used braided and twisted HDPE nets for grow out purpose. It can last for two or more seasons. Nylon net can be used economically, but since it is light weight, to hold the shape intact more weight has to be loaded in the ballast pipe. Cost factor has to be taken care while using new netting materials like sapphire or dyneema materials for net cage. The depth of net ranging from 2 to 5 m is ideal. For open sea cage culture, predator net to prevent attack by predatory organisms is essential.

Potential Species and Criteria for Selection of Species for Cage Culture

The selection of species for cage culture should be based on a number of biological criteria such as omnivore or carnivore, hardiness, fast growing, efficient food conversion ability, availability of quality seeds, disease resistance and market demand.

Stocking

Although stocking densities should be determined by species requirements and operational considerations, the influence of stocking densities on growth and production has been determined empirically. The stocking density depends also on the carrying capacity of the cages and the feeding habits of the cultured species. Optimal stocking density varies with species and size of fish.

Feeds and Feed Management

Fresh or frozen trash fish, moist pellet (MP) and floating dry pellets are the commonly utilized feed for growing fish in cages. Feeding in cages is quite easy compared to that in ponds. The ration can be divided into equal portions and supplied at regular intervals. Feeding can be done either by broadcasting or using feeding trays. Feeds must be nutritionally complete and provide the necessary proteins, carbohydrates, fats, vitamins and minerals needed for growth and health. Feeds cannot be allowed to deteriorate during storage.

Harvest

Harvest of fish in cages is less labour intensive when compared to that in ponds. Floating cages can be towed to a convenient place and full or partial harvest can be carried out based on demand. Marketing of fishes in live conditions as a value addition can also be done.

Cage Management

Cage culture management must result in optimizing production at minimum cost. The management should be so efficient that the cultured fish should grow at the expected rate with respect to feeding rate and stocking density, minimize loss due to disease and predators, monitor environmental parameters and maintain efficiency of the technical facilities. Physical maintenance of cage structures is also of vital importance. The net-cages must be routinely inspected. Necessary repairs and adjustments to anchor ropes and net-cages should be carried out without any delay. Monthly exchange of net should also be considered, as this ensures a good water exchange in the net, thereby washing away faeces, uneaten food and to a certain extent reduce the impact of fouling.

Fouling of Cage Net

Fouling of cage nets and other structures has been observed at many instances of cage farming. Nets get covered with biofoulers. Fouling by molluscs, especially edible oyster and sand barnacles have to be checked before its growth advancement. Algal mats and other periphyton can be removed by introduction of omnivorous grazers in cages. A fouled net will be heavier, thereby increasing drag thus resulting loss of nets and fish.

To avoid/ reduce fouling, net should be changed when required, which may vary from 2 to 4 weeks depending on the intensity of fouling. During oyster fouling, net exchange has to be done immediately after the seasonal spat fall. Herbivorous fish such as rabbit fish (Siganus spp.), pearl spot (Etroplus suratensis) and scat (Scatophagus sp.) can be used to control biofoulers, but their application on a large scale needs to be assessed.

Disease monitoring

Monitoring of fish stock health is essential and early indications can often be observed from changes in behavior, especially during feeding.

Advantages and Disadvantages of Cage Culture

Cage culture of fish has advantages and disadvantages that should be considered carefully before cage production becomes the chosen method. A potential fish farmer can produce fish in an existing pond without destroying sport fishing; does not have to invest large amounts of capital for construction or equipment; and can, therefore, try fish culture without unreasonable risks.

Advantages

Cage culture has advantages which include:

- Many types of water resources can be used, including lakes, reservoirs, ponds, strip pits, streams and rivers which could otherwise not be harvested.
- A relatively low initial investment is required in an existing body of water.
- Harvesting is simplified.
- Observation and sampling of fish is simplified.
- Allows the use of the pond for sport fishing or the culture of other species.
- Less manpower requirement.
- Generation of job opportunities for unemployed youth and women.
- Additional income to fishers during closed seasons.

Disadvantages

Cage culture also has some distinct disadvantages. These include:

- Feed must be nutritionally complete and kept fresh.
- Low Dissolved Oxygen Syndrome (LODOS) is an ever present problem and may require mechanical aeration.
- Fouling of net cage.
- The incidence of disease can be high and diseases may spread rapidly.
- Vandalism or poaching is a potential problem.
- Navigation issues.
- Accumulation of unused feed and excreta will lead to water pollution as well as eutrophication.
- Change in water quality parameters.

- Conflicts within the local community.
- Predation by aquatic mammals and birds.
- Escapement.
- Overcrowding of aquatic organisms in cages.

Copper-alloy Nets

Copper alloy netting improves fish production, leading to healthier fish, higher yields, and lower costs. In addition, copper netting:

- Resists storm damage and lasts longer than traditional netting.
- Reduces predator attacks and fish escapes.
- Stays naturally clean, reduces drag and maintains cage volume.
- Decreases impact of pathogens and parasites, as monitored by farmers.
- Supports sustainable fish farming and is 100% recyclable.
- Minimizes maintenance cost and efforts.

Irrigation Ditch or Pond Systems

These use irrigation ditches or farm ponds to raise fish. The basic requirement is to have a ditch or pond that retains water, possibly with an above-ground irrigation system (many irrigation systems use buried pipes with headers). Using this method, one can store one's water allotment in ponds or ditches, usually lined with bentonite clay. In small systems the fish are often fed commercial fish food, and their waste products can help fertilize the fields. In larger ponds, the pond grows water plants and algae as fish food. Some of the most successful ponds grow introduced strains of plants, as well as introduced strains of fish.

Control of water quality is crucial. Fertilizing, clarifying and pH control of the water can increase yields substantially, as long as eutrophication is prevented and oxygen levels stay high. Yields can be low if the fish grow ill from electrolyte stress.

Integrated Recycling Systems

One of the largest problems with freshwater aquaculture is that it can use a million gallons of water per acre (about 1 m^3 of water per m^2) each year. Extended water purification systems allow for the reuse (recycling) of local water.

The largest-scale pure fish farms use a system derived (admittedly much refined) from the New Alchemy Institute in the 1970s. Basically, large plastic fish tanks are placed in a greenhouse. A hydroponic bed is placed near, above or between them. When tilapia are raised in the tanks, they are able to eat algae, which naturally grows in the tanks when the tanks are properly fertilized.

The tank water is slowly circulated to the hydroponic beds where the tilapia waste feeds a commercial plant crops. Carefully cultured microorganisms in the hydroponic bed convert ammonia to nitrates, and the plants are fertilized by the nitrates and phosphates. Other wastes are strained out by the hydroponic media, which doubles as an aerated pebble-bed filter.

This system, properly tuned, produces more edible protein per unit area than any other. A wide variety of plants can grow well in the hydroponic beds. Most growers concentrate on herbs (e.g. parsley and basil), which command premium prices in small quantities all year long. The most common customers are restaurant wholesalers.

Since the system lives in a greenhouse, it adapts to almost all temperate climates, and may also adapt to tropical climates. The main environmental impact is discharge of water that must be salted to maintain the fishes' electrolyte balance. Current growers use a variety of proprietary tricks to keep fish healthy, reducing their expenses for salt and waste water discharge permits. Some veterinary authorities speculate that ultraviolet ozone disinfectant systems (widely used for ornamental fish) may play a prominent part in keeping the Tilapia healthy with recirculated water.

A number of large, well-capitalized ventures in this area have failed. Managing both the biology and markets is complicated.

Classic Fry Farming

Classic fry farming is also called a "flow through system Trout and other sport fish are often raised from eggs to fry or fingerlings and then trucked to streams and released. Normally, the fry are raised in long, shallow concrete tanks, fed with fresh stream water. The fry receive commercial fish food in pellets. While not as efficient as the New Alchemists' method, it is also far simpler, and has been used for many years to stock streams with sport fish. European eel (Anguilla anguilla) aquaculturalists procure a limited supply of glass eels, juvenile stages of the European eel which swim north from the Sargasso Sea breeding grounds, for their farms. The European eel is threatened with extinction because of the excessive catch of glass eels by Spanish fishermen and overfishing of adult eels in, e.g., the Dutch IJsselmeer, Netherlands. As per 2005, no one has managed to breed the European eel in captivity.

Fishery

A fishery is an organized effort by humans to catch fish or other aquatic species, an activity known as fishing.

Generally, a fishery exists for the purpose of providing human food, although other aims are possible (such as sport or recreational fishing), or obtaining ornamental fish or fish products such as fish oil.

Regardless of purpose, however, the term fishery generally refers to a fishing effort centered on either a particular ecoregion or a particular species or type of fish or aquatic animal, and usually fisheries are differentiated by both criteria.

Examples would be the salmon fishery of Alaska, the cod fishery off the Lofoten islands or the tuna fishery of the Eastern Pacific.

Most fisheries are marine, rather than freshwater; most marine fisheries are based near the coast.

This is not only because harvesting from relatively shallow waters is easier than in the open ocean, but also because fish are much more abundant near the coastal shelf, due to coastal upwelling and the abundance of nutrients available there.

Types of Fishery

There are two main types of fisheries: Inland fisheries and marine fisheries.

Inland or Fresh Water Fisheries

Inland fishery deals with the fishery aspects of waters other than marine water. Potentially, the vast and varied inland fishery resources of India are one of the richest in the world. They pertain to two types of waters, namely, the fresh and the brackish. The former includes the country's great river systems, an extensive network of irrigation canals, reservoirs, lakes, tanks, ponds, etc.

The estuaries, lagoons and mangrove swamps constitute the brackish type of water. In pisciculture (culture fisheries), which generally pertains to small water bodies, the fish seed has to be sown, tended, nursed, reared and finally harvested when grown to table size. In the case of capture fisheries, which pertain to the rivers, estuaries, large reservoirs, as well as big lakes, man has only to reap without having to sow.

	Name		Name
1.	Rohu (Labeo rohita)	5.	Mangur (Clarias batrachus)
2.	Calbasu (L. calbasu)	6.	Singhi (Heteropneustes fossilis)
3.	Catla (Catla catla)	7.	Malli (Wallago attu)
4.	Singhara (Mystus seenghala)	8.	Mirgala (Cirrhinus mrigala)

Types of Breeding

According to the mode of breeding there are two categories, natural breeding and induced breeding.

1. Natural Breeding (Bundh breeding):

The natural bundhs are special types of ponds where natural water resource conditions are managed for the breeding of culturable fish. These bundhs are constructed in large low-lying areas to accumulate large quantity of rain water. These bundhs are having an outlet for the exit of excess rain water.

2. Induced Breeding:

In artificial method of fertilization ova from the females and the sperms from the males are taken out by artificial mechanical process and the eggs are got fertilized by the sperms. Different methods are used for induced breeding. Here induced breeding by hormones method is briefly described. The gonadotropin hormone (FSH and LH) secreted by pituitary gland influences the maturation of gonads and spawning in fishes.

3. Composite Fish Farming:

It is found that if few selected species of fish are stocked together in proper proportion in a pond, total production of fish is increased many times. This mixed farming is called composite farming. It has some advantage-compatible species do not harm each other, all available areas are fully utilised, no competition among different species is found and fish may have beneficial effect on each other. Catla catla, Labeo-rohita and Cirrhina mrigala are surface feeder, column feeder and bottom feeder respectively and are used for composite farming.

Marine Fisheries

Marine fishery deals with the fishery aspects of the sea water or ocean.

Fish Diseases caused by Parasites and Pathogens

1. Bacterial Diseases: Two bacterial diseases are very important.
   - (i) Abdominal dropsy of Carps is caused by Aeromonas punctuata.
   - (ii) Furunculoris of Salmons and trout's is caused by Aeromonas salminicida.
2. Viral Diseases: Economically most important is the viral haemorrhage septicaemia (VHS) of rainbow trouts.
3. Protozoan Diseases: Main protozoan diseases are caused by Costia, Myxobolus and Trypanosoma.
4. Fungal Diseases: The gill rot (branchyomyces) of carps involves the attack of Saprolegnia on the gills of carps.
5. Worm Diseases: Worms of four groups are parasites on fish. The flatworms (trematodes), tapeworms (cestodes), round worms (nematodes) and thorny-headed worms (acanthocephalans).
6. Common Ectoparasites: Two ectoparasites of fish are most important, the fish lice (Argulus, Lernaea and Ergasilus) and the fish leech (Piscicold). Both parasites weaken fish by feeding on its blood.

Fish Farming Issues

Environmental Problems with Fish Farming

Fish farming is a way to create a much larger amount of fish much more quickly, cheaply and efficiently than with wild caught fish. Unfortunately, when something seems to good to be true, it very often is.

Pollution

This density of fish creates problems like disease and pollution. The biggest source of pollution is the accumulation of fish waste and uneaten food beneath the sea pens which can degrade the quality of the surrounding water.

Like commercial farming operations on land, the density of fish in these pens necessitates certain chemicals to keep animals from getting sick and to keep things clean. The chemicals used in marine aquaculture operations such as medicines like antibiotics and vaccines, disinfectants, and substances used to prevent corrosion of equipment (cages, etc.) can also change the composition of the surrounding aquatic ecosystem.

The amount of pollution from fish farms also depends on how the fish are contained. Open-net, or pen systems, allow for a direct exchange of water, where as "closed contentment" methods have a barrier which filters the water.

Impact of Biodiversity

Another way aquaculture can have a negative impact is by introducing farmed species into the wild and therefore changing the biodiversity of aquatic ecosystems. Even when measures are taken to prevent escapes, predators like birds and sharks, equipment failure, human error, severe weather and other complications mean that escapes of farmed fish are inevitable.

Since farmed fish often have been bred via selective breeding they have a lower genetic variation than wild fish. If they interbreed with the wild fish it can result in a less genetically diverse, and therefore less robust, population.

Another concern is infertile offspring. For example, Atlantic and Pacific salmon belong to different genera and while they can produce offspring, those offspring will be unable to reproduce (like mules). If populations of non-native species become established they compete with native populations for resources such as food and breeding sites.

Since farmed fish are selected and bred for certain genetic criteria like size, quick growth and hardiness, escaped species can become invasive, which has been recognized as one of the main causes of global biodiversity loss. One example of this was the Pacific oyster in the UK, which was introduced into its waters in the 1960s via aquaculture with the idea that it would be a more commercially viable species than the native oyster. Unfortunately, these pacific oysters have spread and created reef formations, forcing out the native oysters and altering the marine environment.

Tilapia Takeover

Another example of the negative effect of fish farming on native fish population and environment is with Tilapia. Tilapia is one of the most common types of farmed fish. Most of our tilapia supply is imported from Latin American and Asia, and in 2015, Americans ate 475 million pounds of tilapia.

Tilapia is a warm-water fish native to Africa, but in the last 60 years the governments of poor tropical countries saw the fish as a solution to control weeds and mosquitoes in lakes and rivers, breeding and releasing tilapia into these areas. They are now seen as a nuisance, as they are one of the "most invasive species known and difficult to get rid of once established," says Aaron McNevin.

In Lake Apoyo in Nicaragua, tilapia escaped from a fish farm and their pollution and feeding reduced the lake's quantity of an aquatic plant called charra, which was an important source of food for the lake's native fish populations. Sixteen years later, the lake's biota are still recovering.

Spread of Disease and Antibiotic use

Because farmed fish are raised on unnatural diets and in small enclosures they often breed disease, which can pass to wild populations. This is becoming an increasingly big problem, as are the solutions often used for these diseases.

Some aquaculture productions rely on prophylactic antibiotics to prevent infections. The use of antibiotics can cause drug resistant bacteria to develop which can spread to wild populations.

Sea Lice

Another common disease is sea lice. Not to be confused with an itchy, stinging rash caused by jellyfish larvae, these sea lice are planktonic marine parasites which feed on many types of fish. There are many species but the common "salmon louse" or lepeophtheirus salmonis, has become a big problem for both wild and farmed salmon populations. About a centimeter in size, the sea lice attach themselves to the outside of a fish and feed on its mucous, blood, and skin.

This can cause serious damage to fins, erosion of skin, constant bleeding, and open wounds at risk of infection. On an adult fish this may be only a nuisance, but for small juvenile salmon (around the size of a finger), sea lice can be fatal.

Before offshore industrial scale fisheries became big business in the 1970s, sea lice were rarely epidemic to fish populations. Of course, when hundreds or even thousands of fish are crowded together in a small area, sea lice, and other diseases can easily spread from fish to fish.

This problem not only impacts food supply and fish industry profits, it is spreading to wild fish populations. One example is the salmon in the Broughton Archipelago, a group of islands 260 miles northwest of Vancouver, British Columbia.

In 2007 the area had 20 active fish farms, which raised between 500,000 and 1.5 million fish each. As juvenile wild salmon swam past these open-net farms on their way down river towards the sea, the sea lice infecting the farmed salmon attached to them. A study done that year found that the number of wild pink salmon were down 80% since 1970 because of sea lice infestations. The study concluded that at this rate the wild salmon in the area would die off in four generations or by 2015. While the conclusions of this study were not without controversy, it did seem that the salmon populations recovered when the farms idled.

Pesticide use

As sea lice became a problem in fisheries around the world, an unfortunately common solution was employed: pesticides. One chemical commonly used was emamectin benzoate, or Slice, which when administered to rats and dogs causes tremors, spinal deterioration and muscle atrophy.

Of course soon the lice became resistant, and Slice only worked in triple doses. Other chemicals like hydrogen peroxide, Salmosan, AlphaMax and Calicide chemicals have been employed instead.

While we know that these chemicals can negatively affect ocean water and plant species, we don't have enough research to know how much of these chemicals are absorbed and retained by the fish and if any of this passes to those who eat the fish.

Fish Farming: Effects on Fish

As you might imagine, most species of fish don't thrive when being raised in extremely cramped pens, fed commercial feed, and treated with pesticides, antibiotics and other chemicals. We now know that these modern practices negatively affect the fish as well as their environment.

Higher Levels of Omega-6

Like all animals, fish are what they eat. The nutrition of our food depends on the nutrition of our food's food. For example, salmon in the wild eat smaller fish, which eat aquatic plants rich in beneficial long chain omega-3 fatty acids DHA and EPA. Farm-raised salmon eat pellets, and as the nutritional quality of pellets varies, so does the nutritional quality of the fish. Often young salmon are fed pellets made from plant and animal sources, and they receive more expensive fish/fish oil enriched pellets later in their lifespan just before harvest.

New commercial fish feeds are more likely to have protein and oils derived from grains and oilseeds (like soybeans and canola) and with less fishmeal and fish oil. The difference in feeds accounts for why one study that measured the omega-3 contentment of fish species from six regions of the US found large variations in the omega-3 content in the five salmon species tested.

In the two farm raised varieties tested the omega-3 ranged from 717 mg to 1,533 mg per 100 grams of fish (which is equal to a 3.5 oz serving). Compared to the wild-caught varieties, these farmed fish tended to have higher levels of omega-3s but only because the farmed salmon have more fat overall, including higher levels of omega-6 polyunsaturated fats and saturated fats.

Feeds from vegetable sources can be more sustainable than fishmeal and fish oil. These are often made from smaller fish, lower on the food chain which are sometimes called reduction, pelagic, or trash fish. To create 1kg (2.2 lbs) of fishmeal it takes 4.5 kg (10 lbs) of smaller fish. In fact, today at least 37% of global seafood is ground up to make feed. In 1948 that number was only 7.7%.

These lower food chain fish are the food for many species of aquatic life, and depleting them may cause serious implications for aquatic ecosystems and other sea animals including birds and mammals.

PCBs and POPs

PCBs (polychlorinated biphenys) are industrial pollutants that find their way into fresh waters and oceans and then are absorbed by aquatic wildlife. PCBs are a type of POP (persistent organic pollutant).

Type 2 diabetes and obesity have been linked to POPs, and certain types increase the risk of stroke in women. PCBs are potential human carcinogens, and known to promote cancer in animals. Other potential health effects include negative effects on the reproductive, nervous, and immune systems plus impaired memory and learning.

One study found that PCB levels in farmed salmon, especially those in Europe were five to ten times higher than PCBs in wild salmon. Follow-up studies haven't confirmed this, and there are now strict rules on contaminant levels in feed ingredients which have lowered PCBs in these fish.

It is best to avoid these chemicals completely, but most PCBs are found in the skin, so if farmed raised fish is the only option available, it is possible to reduce exposure by removing the skin and by avoiding fried fish.

Things to Consider when Selecting Seafood

As if the various concerns associated with fish farming weren't enough, there are other important factors to consider when sourcing any kind of seafood.

Mercury

Mercury toxicity can impact brain development in children and negatively affect cognitive function in adults. Mercury is found in the muscle of the fish. The biomagnification of mercury means that organisms higher on the food chain contain higher levels of the metal.

One way to consume seafood yet reduce mercury exposure is by eating smaller fish lower on the food chain, such as sardines.

The Selenium Myth

Many of us in the real-food community have heard that mercury is only a concern if there is not selenium present in the fish, and since most seafood also has high levels of selenium we shouldn't be concerned about mercury.

Dr. Christopher Shade, recently confirmed in an interview with Chris Kresser that this is not the case. He verified that those who are deficient in selenium will be more susceptible to mercury toxicity; however, having good selenium levels doesn't prevent someone from getting mercury toxicity from seafood. Nor does the selenium in seafood bind the mercury and therefore prevent us from absorbing the toxic metal.

One important way the body rids itself of mercury is via glutathione, the body's self-produced master-antioxidant. It is therefore important to support this pathway by consuming sulfur containing vegetables like onions and brassicas, and good amounts of vitamin C.

Omega-3 Levels

Omega-3s are very important for health, and should be consumed in proper ratio with Omega-6 fatty acids. Statistically, most of us consume too much Omega-6 and not enough Omega-3, which some experts blame as one of the root causes of many modern diseases. Fish are an excellent natural source of Omega-3s, but there is a wide range of levels depending on the fish. When choosing seafood, it helps to know which fish have the highest levels of these beneficial fats.

Sustainability

While farmed fish have obvious drawbacks there are also sustainability concerns about wild caught fish.

One of the major concerns is overfishing which has become a global problem. Obviously it becomes difficult to eat the fish if they don't exist, but fewer populations of certain species can have repercussions for an entire ecosystem.

Another issue is bycatch, which is when non-target animals are caught during fishing. This can include dolphins, sea turtles, birds, sharks, stingrays, and other fish like juvenile fish.The incidence of bycatch can be reduced by the use of selective fishing gear designed to catch only the species selected and implementing measures to return the native species.

Habitat destruction can degrade aquatic ecosystems, as seabed habitats provide shelter and food for a variety of species. One fishing method that is a common culprit is bottom trawling near vulnerable areas like coral reefs or breeding and nursing grounds.

Indoor Fish Farming

An alternative to outdoor open ocean cage aquaculture, is through the use of a recirculating aquaculture system (RAS). A RAS is a series of culture tanks and filters where water is continuously recycled and monitored to keep optimal conditions year round. To prevent the deterioration of water quality, the water is treated mechanically through the removal of particulate matter and biologically through the conversion of harmful accumulated chemicals into nontoxic ones.

Other treatments such as ultraviolet sterilization, ozonation, and oxygen injection are also used to maintain optimal water quality. Through this system, many of the environmental drawbacks of aquaculture are minimized including escaped fish, water usage, and the introduction of pollutants. The practices also increased feed-use efficiency growth by providing optimum water quality.

Top four myths about indoor fish farming are:

- Indoor fish farming hurts the environment:

  Environmentalists have good reason to be suspicious of fish farms. Previously, these ocean-bound fish stables have been breeding grounds for diseases that can spread to nearby wild marine life. But, indoor fish farming involving recirculating aquaculture systems (RAS), are actually very sustainable. Indoor fish farming is often considered environmentally friendly because it requires less water and produces less waste. Many systems are rather sophisticated and allow for automatic collection and processing of fish wastes into crop fertilizers. As for the spread of disease, these fish only interact with each other in stabilized water, so disease isn't introduced or passed on.

On land, indoor fisheries can be built closer to cities, so less transportation and fossil fuels are required from the farm to the market. Fish grown in the US for US consumers do not involve shipping frozen fish over long distances. On the other hand, fishing boats and the necessary supply-chain distribution of wild-caught fish has a much higher carbon-footprint.

- Indoor fish farming raises the price for consumption:

On the economic side, farm-raised fish are bred to make fish cheaper and more readily available to consumers. Currently, America's aquaculture industry supports only 6% of US food demand, producing primarily oysters, clams, mussels, and some fish.

About 90% of fish consumed by Americans is imported, increasing fish prices and contributing to the country's trade deficit. Additionally, global production of fish is much more inconsistent than indoor raised fish, and is susceptible to weather patterns and biological factors (disease, predators). As a result, prices to distributors fluctuate much more than wild caught fish.

- Fish raised indoors isn't nutritious or tasty:

Farm-raised fish have more omegas than fish raised in the wild, due to their higher fat content. Farmed salmon, in many cases, can be just as nutritious as its wild-caught salmon, and can be richer in omega-3s and omega-6s essential fatty acids. As for texture, farmed fish tend to have a little bit more fat in their diet, so they might be a little more tender or softer, compared to a wild-caught fish which may be a little leaner.

According to the Seattle Times, farm-raised salmon had a superior taste profile compared to wild caught salmon. 10 different types of frozen salmon ranging from wild-caught in Alaska to farmed in Norway, were prepared by a head chef and served to tasters from The Washington Post Food section and the D.C. area seafood scene. The panel ranked Costco's frozen farmed salmon the best for taste and texture. The bottom of the list was Costco's frozen wild-caught salmon. Everything else fell between, but farmed-fish took the top five spots, and all the wild-caught rounded out the bottom.

- Farm-raised fish is dangerous to eat because of chemicals and pesticides:

As a rule, RAS operations do not normally utilize antibiotics, except under the rare cases where poor management practices causes where disease to be introduced into a specific tank. The chance of this happening is minimized compared to fish raised in outdoor pins, as inputs and outputs are tightly controlled. In any event, were a specific batch of fish to be treated, like in all animal husbandry, the fish would not be sent to market until a certain period had passed and the antibiotics were out of their system.

And unlike wild-caught fish, indoor grown fish do not pose a threat of mercury poisoning, an increasing problem in fishes harvested in some areas of the world. Toxic methylmercury from the coal-burning power plants that rim the Northern Pacific Ocean in the US, Japan, China, and Mexico can enter the bodies of large wild caught fish, while fish hatched and raised indoors do not face this threat.

Recirculating Aquaculture Systems

Recirculation aquaculture is essentially a technology for farming fish or other aquatic organisms by reusing the water in the production. The technology is based on the use of mechanical and biological filters, and the method can in principle be used for any species grown in aquaculture such as fish, shrimps, clams, etc. Recirculation technology is however primarily used in fish farming.

Recirculation is growing rapidly in many areas of the fish farming sector, and systems are deployed in production units that vary from huge plants generating many tonnes of fish per year for consumption to small sophisticated systems used for restocking or to save endangered species.

Recirculation can be carried out at different intensities depending on how much water is recirculated or re-used. Some farms are super intensive farming systems installed inside a closed insulated building using as little as 300 litres of new water, and sometimes even less, per kilo of fish produced per year. Other systems are traditional outdoor farms that have been rebuilt into recirculated systems using around 3 m^3 new water per kilo of fish produced per year. A traditional flowthrough system for trout will typically use around 30 m^3 per kilo of fish produced per year. As an example, on a fish farm producing 500 tonnes of fish per year, the use of new water in the examples given will be 17 m^3 /hour(h), 171 m^3 /h and 1 712 m^3 /h respectively, which is a huge difference.

The formula has been used for calculating the degree of recirculation at different system intensities and also compared to other ways of measuring the rate of recirculation.

Type of system	Consumption of new water per kg fish produced per year	Consumption of new water per cubic meter per hour	Consumption of new water per day of total system water volume	Degree of recirculation at system vol. recycled one time per hour
Flow-through	30 m^3	1 712 m^3 /h	1 028 %	0 %
RAS low level	3 m^3	171 m^3 /h	103 %	95.9 %
RAS intensive	1 m^3	57 m^3 /h	34 %	98.6 %
RAS super intensive	0.3 m^3	17 m^3 /h	6 %	99.6 %

Seen from an environmental point of view, the limited amount of water used in recirculation is of course beneficial as water has become a limited resource in many regions. Also, the limited use of water makes it much easier and cheaper to remove the nutrients excreted from the fish as the volume of discharged water is much lower than that discharged from a traditional fish farm. Recirculation aquaculture can therefore be considered a most environmentally friendly way of producing fish at a commercially viable level. The nutrients from the farmed fish can be used as fertilizer on agricultural farming land or as a basis for biogas production.

The term "zero-discharge" is sometimes used in connection to fish farming, and although it is possible to avoid all discharge from the farm of all sludge and water, the waste water treatment of the very last concentrations is most often a costly affair to clean off completely. Thus an application for discharging nutrients and water should always be part of the planning permission application.

Most interesting though, is the fact that the limited use of water gives a huge benefit to the production inside the fish farm. Traditional fish farming is totally dependent on external conditions such as the water temperature of the river, cleanliness of the water, oxygen levels, or weed and leaves drifting downstream and blocking the inlet screens, etc. In a recirculated system these external factors are eliminated either completely or partly, depending on the degree of recirculation and the construction of the plant.

Recirculation enables the fish farmer to completely control all the parameters in the production, and the skills of the farmer to operate the recirculation system itself becomes just as important as his ability to take care of the fish.

Controlling parameters such as water temperature, oxygen levels, or daylight for that matter, gives stable and optimal conditions for the fish, which again gives less stress and better growth. These stable conditions result in a steady and foreseeable growth pattern that enables the farmer to precisely predict when the fish will have reached a certain stage or size. The major advantage of this feature is that a precise production plan can be drawn up and that the exact time the fish will be ready for sale can be predicted. This favours the overall management of the farm and strengthens the ability to market the fish in a competitive way.

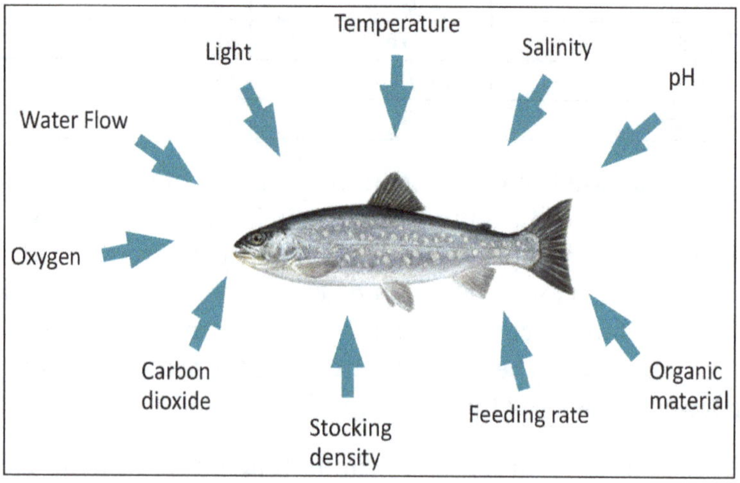

There are many more advantages of using recirculation technology in fish farming. However, one major aspect to be mentioned right away is that of diseases. The impact of pathogens is lowered considerably in a recirculation system as invasive diseases from the outside environment are minimised by the limited use of water. Water for traditional fish farming is taken from a river, a lake or the sea, which naturally increases the risk of dragging in diseases. Due to the limited use of water in recirculation the water is mainly taken from a borehole, drainage system or spring where the risk of diseases is minimal. In fact, many recirculation systems do not have any problems with diseases whatsoever, and the use of medicine is therefore reduced significantly for the benefit of the production and the environment. To reach this level farming practice it is of course extremely important that the fish farmer is very careful about the eggs or fry that he brings on to his farm. Many diseases are carried into systems by taking in infested eggs or fish for stocking. The best way to avoid diseases entering this way, is not to bring in fish from outside, but only bring in eggs as these can be disinfected completely from diseases.

Aquaculture requires knowledge, good husbandry, persistence and sometimes nerves of steel. Shifting from traditional fish farming into recirculation does make many things easier, however at the same time it requires new and greater skills.

Recirculation System

In a recirculation system it is necessary to treat the water continuously to remove the waste products excreted by the fish, and to add oxygen to keep the fish alive and well. A recirculation system is in fact quite simple. From the outlet of the fish tanks the water flows to a mechanical filter and further on to a biological filter before it is aerated and stripped of carbon dioxide and returned to the fish tanks. This is the basic principle of recirculation.

Several other facilities can be added, such as oxygenation with pure oxygen, ultraviolet light or ozone disinfection, automatic pH regulation, heat exchanging, denitrification, etc. depending on the exact requirements.

The basic water treatment system consists of mechanical filtration, biological treatment and aeration/ stripping. Further installations, such as oxygen enrichment or UV disinfection, can be added depending on the requirements.

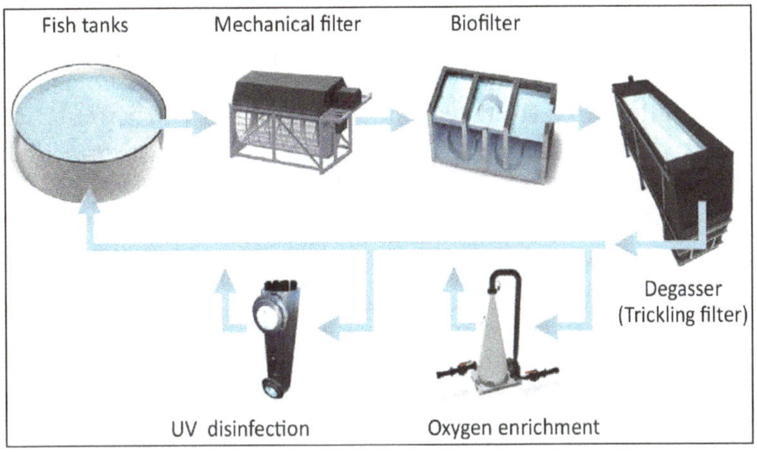

Figure: Principle drawing of a recirculation system.

Fish in a fish farm require feeding several times a day. The feed is eaten and digested by the fish and is used in the fish metabolism supplying energy and nourishment for growth and other physiological processes. Oxygen (O_2) enters through the gills, and is needed to produce energy and to break down protein, whereby carbon dioxide (CO_2) and ammonia (NH_3) are produced as waste products. Undigested feed is excreted into the water as faeces, termed suspended solids (SS) and organic matter. Carbon dioxide and ammonia are excreted from the gills into the water. Thus fish consume oxygen and feed, and as a result the water in the system is polluted with faeces, carbon dioxide and ammonia.

Figure: Eating feed and using oxygen results in fish growth and excretion of waste products, such as carbon dioxide, ammonia and faeces.

Only dry feed can be recommended for use in a recirculation system. The use of trash fish in any form must be avoided as it will pollute the system heavily and infection with diseases is very likely. The use of dry feed is safe and also has the advantage of being designed to meet the exact biological needs of the fish. Dry feed is delivered in different pellet sizes suitable for any fish stage, and the ingredients in dry fish feed can be combined to develop special feeds for fry, brood stock, grow-out, etc.

In a recirculation system, a high utilization rate of the feed is beneficial as this will minimise the amount of excretion products thus lowering the impact on the water treatment system. In a professionally managed system, all the feed added will be eaten keeping the amount of uneaten feed to a minimum. The feed conversion rate (FCR), describing how many kilos of feed you use for every kilo of fish you produce, is improved, and the farmer gets a higher production yield and a lower impact on the filter system. Uneaten feed is a waste of money and results in an unnecessary load on the filter system. It should be noted that feeds especially suitable for use in recirculation systems are available. The composition of such feeds aims at maximising the uptake of protein in the fish thus minimising the excretion of ammonia into the water.

Table: Ingredients and content of a trout feed suitable for use in a recirculation system.

Pellet size	Fish size, gram	Protein	Fat
3 mm	40 – 125	43%	27 %
4.5 mm	100 – 500	42 %	28 %
6.5 mm	400 – 1200	41 %	29 %

Composition, %	3.0 mm	4.5 mm	6.5 mm
Fishmeal	22	21	20
Fish oil	9	10	10
Rape seed oil	15	15	16
Haemoglobin meal	11	11	11
Peas	5	5	5
Soya	10	11	11
Wheat	12	11	11
Wheat gluten	5	5	5
Other protein concentrates	10	10	10
Vitamins, minerals, etc.	1	1	1

Components in a Recirculation System

Fish Tanks

Table: Different tank designs give different properties and advantages. Rating 1-5, where 5 is the best.

Tank properties	Circular tank	D-ended raceway	Raceway type
Self-cleaning effect	5	4	3
Low residence time of particles	5	4	3
Oxygen control and regulation	5	5	4
Space utilization	2	4	5

The environment in the fish rearing tank must meet the needs of the fish, both in respect of water quality and tank design. Choosing the right tank design, such as size and shape, water depth, self-cleaning ability, etc. can have a considerable impact on the performance of the species reared.

If the fish is bottom dwelling, the need for tank surface area is most important, and the depth of water and the speed of the water current can be lowered (turbot, sole or other flatfish), whereas pelagic living species such as salmonids will benefit from larger water volumes and show improved performance at higher speeds of water.

In a circular tank, or in a square tank with cut corners, the water moves in a circular pattern making the whole water column of the tank move around the centre. The organic particles have a relatively short residence time of a few minutes, depending on tank size, due to this hydraulic pattern that gives a self- cleaning effect. A vertical inlet with horizontal adjustment is an efficient way of controlling the current in such tanks.

In a raceway the hydraulics have no positive effect on the removal of the particles. On the other hand, if a fish tank is stocked efficiently with fish, the self-cleaning effect of the tank design will depend more on the fish activity than on the tank design. The inclination of the tank bottom has little or no influence on the self-cleaning effect, but it will make complete draining easier when the tank is emptied.

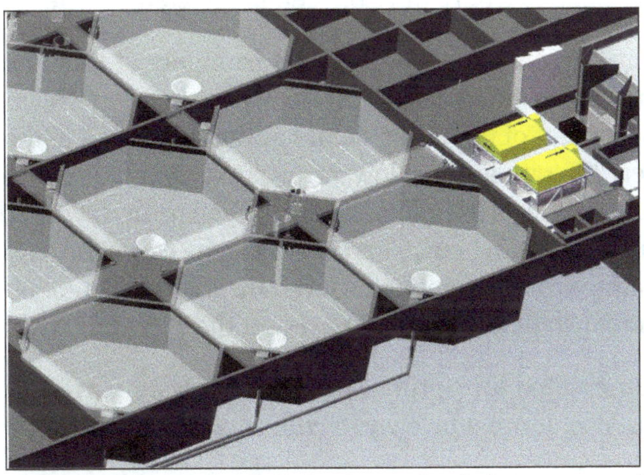

Figure: An example of octagonal tank design in a recirculation system saving space yet achieving the good hydraulic effects of the circular tank.

Circular tanks take up more space compared to raceways, which adds to the cost of constructing a building. By cutting off the corners of a square tank an octagonal tank design appears, which will give better space utilization than circular tanks, and at the same time the positive hydraulic effects of the circular tank are achieved. It is important to note that construction of large tanks will always favour the circular tank as this is the strongest design and the cheapest way of making a tank.

A hybrid tank type between the circular tank and the raceway called a "D-ended raceway" also combines the self-cleaning effect of the circular tank with the efficient space utilization of the raceway. However, in practice this type of tank is seldom used, presumably because the installation of the tank requires extra work and new routines in management.

Sufficient oxygen levels for fish welfare are important in fish farming and are usually kept high by increasing the oxygen level in the inlet water to the tank.

Figure: Circular tank, D-ended raceway, and raceway type.

Direct injection of pure oxygen in the tank by the use of diffusers can also be used, but the efficiency is lower and more costly.

Control and regulation of oxygen levels in circular tanks or similar is relatively easy because the water column is constantly mixed making the oxygen content almost the same anywhere in the tank. This means that it is quite easy to keep the desired oxygen level in the tank. An oxygen probe placed near the tank outlet will give a good indication of the oxygen available. The time it takes for the probe to register the effect of oxygen being added to a circular tank will be relatively short. The probe must not be placed close to where pure oxygen is injected or where oxygen rich water is fed.

In a raceway, however, the oxygen content will always be higher at the inlet and lower at the outlet, which also gives a different environment depending on where each fish is swimming. The oxygen probe for measuring the oxygen content of the water should always be placed in the area with the lowest oxygen content, which is near the outlet. This downstream oxygen gradient will make the regulation of oxygen more difficult as the time lag from adjusting the oxygen up or down at the inlet to the time this is measured at the outlet can be up to an hour. This situation may cause the oxygen to go up and down all the time instead of fluctuating around the selected level. Installation of modern oxygen control systems using algorithms and time constants will however prevent these unwanted fluctuations.

Tank outlets must be constructed for optimal removal of waste particles, and fitted with screens with suitable mesh sizes. Also, it must be easy to collect dead fish during the daily work routines.

Tanks are often fitted with sensors for water level, oxygen content and temperature for having complete control of the farm. It should also be considered to install diffusers for supplying oxygen directly into each tank in case of an emergency situation.

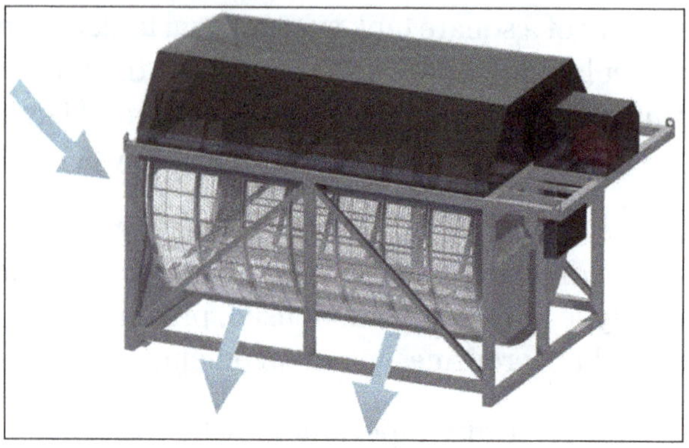

Figure: Drumfilter.

Mechanical Filtration

Mechanical filtration of the outlet water from the fish tanks has proven to be the only practical solution for removal of the organic waste products. Today almost all recirculated fish farms filter the outlet water from the tanks in a so called microscreen fitted with a filter cloth of typically 40 to 100 microns. The drumfilter is by far the most commonly used type of microscreen, and the design ensures the gentle removal of particles.

Function of the Drumfilter

- Water to be filtered enters the drum.
- The water is filtered through the drum's filter elements. The difference in water level inside/outside the drum is the driving force for the filtration.
- Solids are trapped on the filter elements and lifted to the backwash area by the rotation of the drum.
- Water from rinse nozzles is sprayed from the outside of the filter elements. The rejected organic material is washed out of the filter elements into the sludge tray.
- The sludge flows together with water by gravity out of the filter escaping the fish farm for external waste water treatment.

Microscreen filtration has the following advantages:

- Reduction of the organic load of the biofilter.
- Making the water clearer as organic particles are removed from the water.
- Improving conditions for nitrification as the biofilter does not clog.
- Stabilising effect on the biofiltration processes.

Biological Treatment

Not all the organic matter is removed in the mechanical filter, the finest particles will pass through together with dissolved compounds such as phosphate and nitrogen. Phosphate is an inert substance, with no toxic effect, but nitrogen in the form of free ammonia (NH_3) is toxic, and needs to be transformed in the biofilter to harmless nitrate. The breakdown of organic matter and ammonia is a biological process carried out by bacteria in the biofilter. Heterotrophic bacteria oxidise the organic matter by consuming oxygen and producing carbon dioxide, ammonia and sludge. Nitrifying bacteria convert ammonia into nitrite and finally to nitrate.

The efficiency of biofiltration depends primarily on:

- The water temperature in the system.
- The pH level in the system.

To reach an acceptable nitrification rate, water temperatures should be kept within 10 to 35°C (optimum around 30°C) and pH levels between 7 and 8. The water temperature will most often depend on the species reared, and is as such not adjusted to reach the most optimal nitrification rate, but to give optimal levels for fish growth. Regulation of pH in relation to biofilter efficiency is however important as lower pH level reduces the efficiency of the biofilter. The pH should therefore be kept above 7 in order to reach a high rate of bacterial nitrifying. On the other hand, increasing pH will result in an increasing amount of free ammonia (NH_3), which will enhance the toxic effect. The aim is therefore to find the balance between these two opposite aims of adjusting the pH. A recommended adjustment point is between pH 7.0 and pH 7.5.

Two major factors affect the pH in the water recirculation system:

- The production of CO_2 from the fish and from the biological activity of the biofilter.
- The acid produced from the nitrification process.

Result of nitrification process:

$$NH_4\,(\text{ammonium}) + 1.5\ O_2 \longrightarrow NO_2\,(\text{nitrite}) + H_2O + 2H^+ + 2e$$
$$NO_2\,(\text{nitrite}) + 0.5\ O_2 \longrightarrow NO_3\,(\text{nitrate}) + e$$
$$NH_4 + 2O_2 \quad \text{«} NO_3 + H_2O + 2H$$

CO_2 is removed by aeration of the water, whereby degassing takes place. This process can be accomplished in several ways.

The nitrifying process produces acid (H+) and the pH level falls. In order to stabilize the pH, a base must be added. For this purpose lime or sodium hydroxide (NaOH) or another base needs to be added to the water.

Fish excretes a mixture of ammonia and ammonium (Total Ammonia Nitrate (TAN):

= ammonium ($NH_4^{\ +}$) + ammonia (NH_3) where ammonia constitutes the main part of the excretion. The amount of ammonia in the water depends however on the pH level as can be seen in figure, which shows the equilibrium between ammonia (NH_3) and ammonium ($NH_4^{\ +}$).

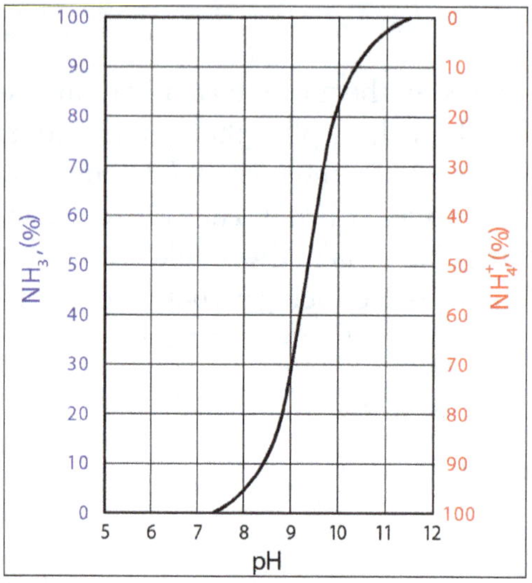

Figure: The equilibrium between ammonia (NH_3) and ammonium (NH_4) at 20°C.
The toxic ammonia is absent at pH below 7, but rises fast as pH is increased.

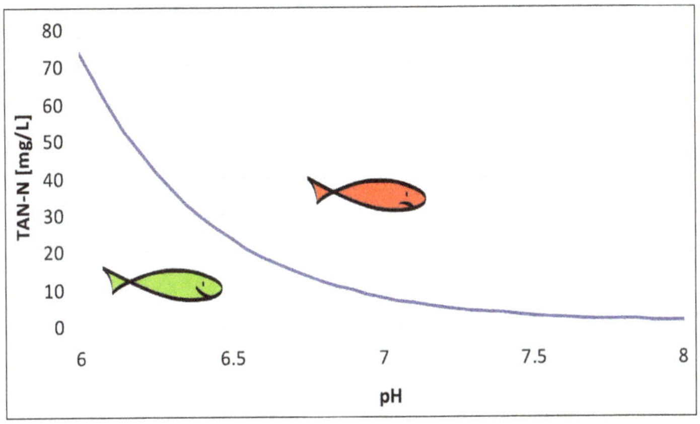

Figure: The relation between measured pH and the amount of TAN available for breakdown in the biofilter, based upon a toxic ammonia concentration of 0.02 mg/L.

In general, ammonia is toxic to fish at levels above 0.02 mg/L. Figure above shows the maximum concentration of TAN to be allowed at different pH levels if a level below 0.02 mg/L of ammonia is to be ensured. The lower pH levels minimises the risk of exceeding this toxic ammonia limit of 0.02 mg/L, but the fish farmer is recommended to reach a level of minimum pH 7 in order to reach a higher biofilter efficiency as explained earlier. Unfortunately, the total concentration of TAN to be allowed is thereby significantly reduced as can be seen in figure. Thus there are two opposite working vectors of the pH that the fish farmer has to take into consideration when tuning his biofilter.

Nitrite (NO_2) is formed at the intermediate step in the nitrification process, and is toxic to fish at levels above 2.0 mg/L. If fish in a recirculation system are gasping for air, although the oxygen concentration is fine, a high nitrite concentration may be the cause. At high concentrations, nitrite is transported over the gills into the fish blood, where it obstructs the oxygen uptake. By adding salt to the water, reaching as little as 0.3 ‰, the uptake of nitrite is inhibited.

Nitrate (NO_3) is the end-product of the nitrification process, and although it is considered harmless, high levels (above 100 mg/L) seem to have a negative impact on growth and feed conversion. If the exchange of new water in the system is kept very low, nitrate will accumulate, and unacceptable levels will be reached. One way to avoid the accumulation is to increase the exchange of new water, whereby the high concentration is diluted to a lower and trouble-free level.

On the other hand, the whole idea of recirculation is saving water, and in some instances water saving is a major goal. Under such circumstances, nitrate concentrations can be reduced by de-nitrification. Under normal conditions, a water consumption of more than 300 litres per kg feed used is sufficient to dilute the nitrate concentration. Using less water than 300 litres per kg feed makes the use of denitrification worth considering.

The most predominant denitrifying bacteria are called Pseudomonas. This is an anaerobic (no oxygen) process reducing nitrate to atmospheric nitrogen. In fact, this process removes nitrogen from the water into the atmosphere, whereby the load of nitrogen into the surrounding environment is reduced. The process requires an organic source (carbon), for example wood alcohol (methanol) that can be added to a denitrification chamber. In practical terms 2.5 kg of methanol is needed for each kg nitrate (NO_3 -N) denitrified.

Most often the denitrification chamber is fitted with biofilter media designed with a residence time of 2-4 hours. The flow must be controlled to keep outlet oxygen concentration at app. 1 mg/L. If oxygen is completely depleted extensive production of hydrogen sulphide (H_2S) will take place, which is extremely toxic to fish and also bad smelling (rotten egg). Resulting production of sludge is quite high, and the unit has to be back-washed, typically once a week.

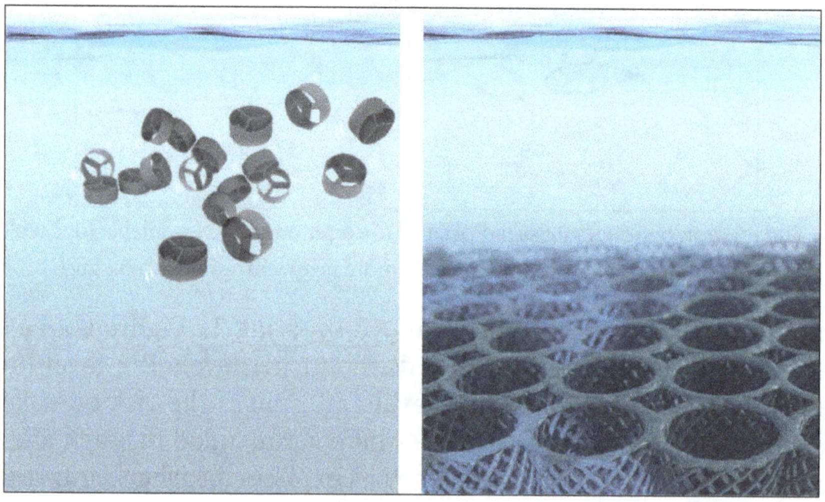

Figure: Moving bed media on left and fixed bed media on right.

Biofilters are typically constructed using plastic media giving a high surface area per m3 of biofilter. The bacteria will grow as a thin film on the media thereby occupying an extremely large surface area. The aim of a well-designed biofilter is to reach as high a surface area as possible per m3 without packing the biofilter so tight that it will get clogged with organic matter under operation. It is therefore important to have a high percentage of free space for the water to pass through and to have a good overall flow through the biofilter together with a sufficient back-wash procedure. Such back-wash procedures must be carried out at sufficient intervals once a week or month depending on the load on the filter. Compressed air is used to create turbulence in the filter whereby organic matter is ripped off. The biofilter is shunted while the washing procedure takes place, and the dirty water in the filter is drained off and discharged before the biofilter is connected to the system again.

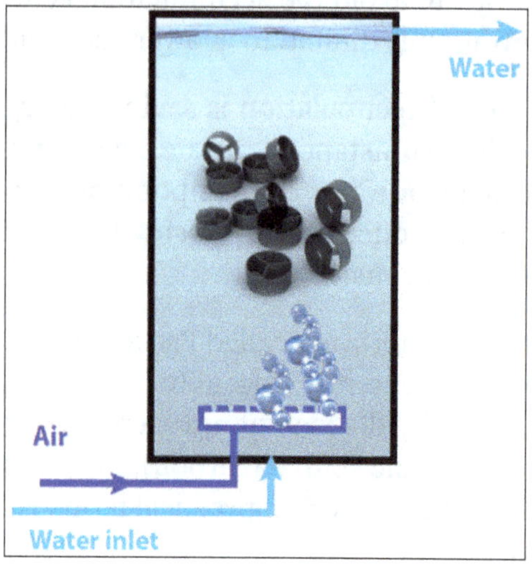

Biofilters used in recirculation systems can be designed as fixed bed filters or moving bed filters. All biofilters used in recirculation today work as submerged units under water. In the fixed bed filter, the plastic media is fixed and not moving. The water runs through the media as a laminar flow to make contact with the bacterial film. In the moving bed filter, the plastic media is moving around in the water inside the biofilter by a current created by pumping in air. Because of the constant movement of the media, moving bed filters can be packed harder than fixed bed filters thus reaching a higher turnover rate per m^3 of biofilter. There is however no significant difference in the turnover rate calculated per m^2 (filter surface area) as the efficiency of the bacterial film in either of the two types of filter is more or less the same. In the fixed bed filter, however, fine organic particles are also removed as these substances adhere to the bacterial film. The fixed bed filter will therefore act also as a fine mechanical filtration unit removing microscopic organic material and leaving the water very clear. The moving bed filter will not have the same effect as the constant turbulence of water will make any adhesion impossible.

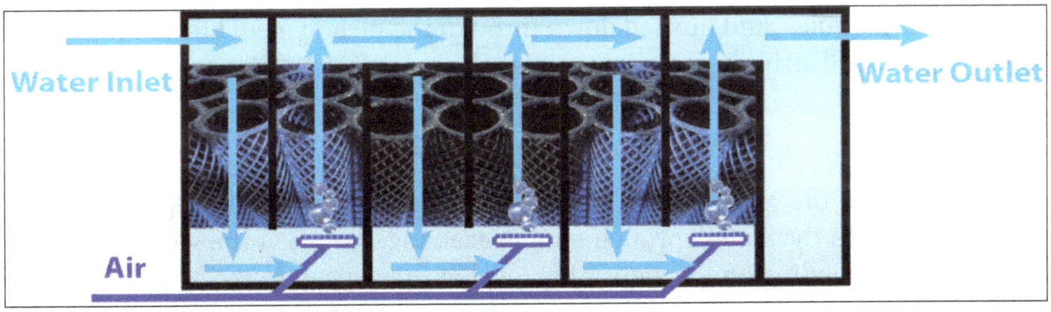

Both filter systems can be used in the same system, or they can be combined; using the moving bed to save space and the fixed bed to benefit from the adhering effect. There are several solutions for the final design of biofilter systems depending on farm size, species to be cultured, sizes of fish, etc.

Degassing, Aeration and Stripping

Before the water runs back to the fish tanks accumulated gases, which are detrimental to the fish, must be removed. This degassing process is carried out by aeration of the water, and the method is often referred to as stripping.

The water contains carbon dioxide (CO_2) from the fish respiration and from the bacteria in the biofilter in the highest concentrations, but free nitrogen (N_2) is also present. Accumulation of carbon dioxide and nitrogen gas levels will have detrimental effects on fish welfare and growth. Under anaerobic conditions hydrogen sulphide (H_2S) can be produced, especially in saltwater systems. This gas is extremely toxic to fish, even in low concentrations, and fish will be killed if the hydrogen sulphide is generated in the system.

Aeration can be accomplished by pumping air into the water whereby the turbulent contact between the air bubbles and the water drives out the gases. This underwater aeration makes it possible to move the water at the same time, for example if an aeration well system is used.

Figure: Trickling filter wrapped in a blue plastic liner to eliminate splashing on the floor. The aeration/stripping process is also called CO_2-stripping. The media in the trickling filter typically consists of the same type of media as used in fixed bed biofilters.

The aeration well system is however not as efficient for removing gases as the trickling filter system, also called a degasser. In the trickling system, gases are stripped off by physical contact between the water and plastic media stacked in a column. Water is led to the top of the filter over a distribution plate with holes, and flushed down through the plastic media to maximise turbulence and contact, the so called stripping process.

Oxygenation

The aeration process of the water, which is the same physical process as degassing or stripping, will add some oxygen to the water through simple exchange between the gases in the water and the gases in the air depending on the saturation level of the oxygen in the water. The equilibrium of oxygen in water is 100% saturation. When the water has been through the fish tanks, the oxygen content has been lowered, typically down to 70%, and the content is reduced further in the biofilter. Aeration of this water will typically bring the saturation up to around 90%, in some systems 100% can be reached. Oxygen saturation higher than 100% in the inlet water to the fish tanks is however often preferred in order to have sufficient oxygen available for a high and stable fish growth. Saturation levels above 100% call for a system using pure oxygen.

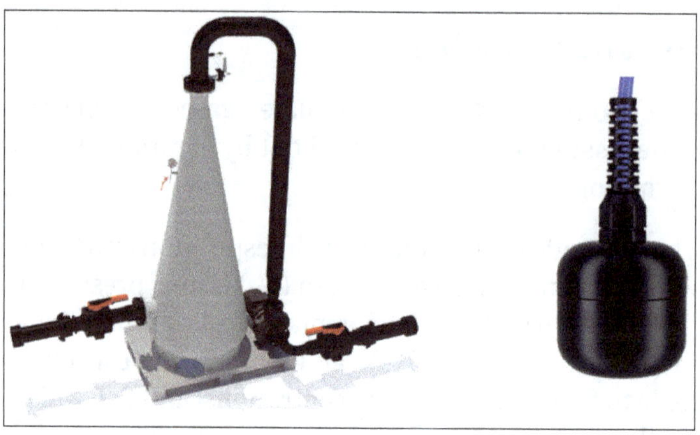

Figure: Oxygen cone for dissolving pure oxygen at high pressure and a sensor (probe) for measuring the oxygen saturati on of the water.

Pure oxygen is often delivered in tanks in the form of liquid oxygen, but can also be produced on the farm in an oxygen generator. There are several ways of making super-saturated water with oxygen contents reaching 200-300 %. Typically high pressure oxygen cone systems or low head oxygen systems, such as oxygen platforms are used. The principle is the same. Water and pure oxygen are mixed under pressure whereby the oxygen is forced into the water. In the oxygen cone the pressure is accomplished with a pump creating a high pressure of typically around 1.4 bar in the cone. Pumping water under pressure into the oxygen cone consumes a lot of electricity. In the oxygen platform the pressure is much lower, typically down to about 0.1 bar, and water is simply pumped through the box mixing water and oxygen. The difference in the two kinds of systems is that the oxygen cone solution uses only a part of the circulating water for oxygen enrichment, whereas the oxygen platform is used for the main recirculation flow often in combination with the overall pumping of water round in the system.

Figure: Oxygen platform for dissolving pure oxygen at low pressure while pumping water around in the farm. The system typically increases the level dissolved oxygen to just above 100% when entering the fish tanks depending on flow rates and farm design.

Whatever method is used, the process should be controlled with the help of oxygen measurement. The best way of doing so is to have the oxygen probe measuring after the oxygenation system at normal atmospheric pressure, for example in a measurement chamber delivered by the supplier. This makes the measurement easier than if it was made under pressure, since the probe will need to be wiped clean and calibrated, from time to time.

Ultraviolet Light

UV disinfection works by applying light in wavelengths that destroy DNA in biological organisms. In aquaculture pathogenic bacteria and one-celled organisms are targeted. The treatment has been used for medical purposes for decades and does not impact the fish as UV treatment of the water is applied outside the fish production area. It is important to understand that bacteria grow so rapidly in organic matter that controlling bacterial numbers in traditional fish farms has limited effect. The best control is achieved when effective mechanical filtration is combined with a thorough biofiltration to effectively remove organic matter from the process water, thus making the UV radiation work efficiently.

The UV dose can be expressed in several different units. One of the most widely used is micro Watt-seconds per cm^2 (μWs/cm^2). The efficiency depends on the size and species of the target organisms and the turbidity of the water. In order to control bacteria and viruses the water needs to be treated with roughly 2 000 to 10 000 μWs/cm^2 to kill 90% of the organisms, fungi will need 10 000 to 100 000 and small parasites 50 000 to 200 000 μWs/cm^2.

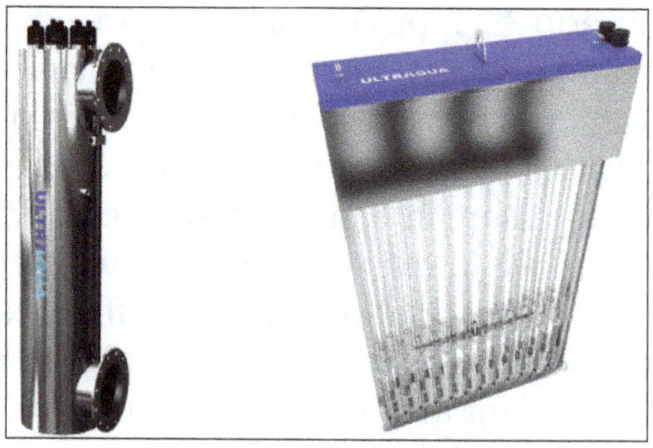

Figure: Closed and open UV treatment systems: For installation in a closed piping system and in an open channel system respectively.

UV lighting used in aquaculture must work under water to give maximum efficiency, lamps fitted outside the water will have little or no effect because of water surface reflection.

Ozone

The use of ozone (O_3) in fish farming has been criticised because the effect of over-dosing can cause severe injury to the fish. In farms inside buildings, ozone can also be harmful to the people working in the area as they may inhale too much ozone. Thus correct dosing and monitoring of the loading together with proper ventilation is crucial to reach a positive and safe result.

Ozone treatment is an efficient way of destroying unwanted organisms by the heavy oxidation of organic matter and biological organisms. In ozone treatment technology micro particles are broken down into molecular structures that will bind together again and form larger particles. By this form of flocculation, microscopic suspended solids too small to be caught can now be removed from the system instead of passing through the different types of filters in the recirculation system. This technology is also referred to as water polishing as it makes the water clearer and free of any suspended solids and possible bacteria adhering to these. This is especially suitable in hatchery and fry systems growing small fish, which are sensitive to micro particles and bacteria in the water.

Ozone treatment can also be used when the intake water to a recirculation system needs to be disinfected.

It is worth mentioning that in many cases UV treatment is a good and safe alternative to ozone.

pH Regulation

The nitrifying process in the biofilter produces acid, thus the pH level will drop. In order to keep a stable pH a base must be added to the water. In some systems a lime mixing station is installed dripping limewater into the system and thereby stabilizing pH. An automatic dosage system regulated by a pH-meter with a feedback impulse to a dosage pump is another option. With this system it is preferable to use sodium hydroxide (NaOH) as it is easy to handle and making the system easier to maintain. Sodium hydroxide is a strong alkaline that can severely burn eyes and skin. Safety precautions must be taken, and glasses and gloves must be worn while handling this and other strong acids and bases.

Water Temperature Regulation

Maintaininganoptimalwatertemperatureintheculturesystemismostimportant as the growth rate of the fish is directly related to the water temperature. Using the intake water is a fairly simple way of regulating the temperature from day to day. In an indoor recirculation system the heat will slowly build up in the water, because energy in the form of heat is released from the fish metabolism and the bacterial activity in the biofilter. Heat from friction in the pumps and the use of other installations will also accumulate. High temperatures in the system are therefore often a problem in an intensive recirculation system. By adjusting the amount of cool fresh intake water into the system, the temperature can be regulated in a simple way.

If cooling by the use of intake water is limited a heat pump can be used. The heat pump will utilize the amount of energy normally lost in the discharge water or in the air leaving the farm. The energy is then used for cooling the circulating water inside the farm. A similar way of lowering heating/cooling cost can be achieved by recovering the energy by the use of a heat exchanger. Energy in the discharge water from the farm is transferred to the cold incoming intake water or vice versa. This is done by passing both streams into the heat exchanger where the warm outlet water will lose energy and heat up the cold intake water, without mixing the two streams. Also on the ventilation system a heat exchanger for air can be mounted utilizing energy from the out-going air and transferring it to the in-going air, thereby reducing the need for heating significantly.

In cold climates heating of the water can be necessary. The heat can come from any source like an oil or gas boiler and is, independent of energy source, connected to a heat exchanger to heat the recirculated water. Heat pumps are an environmental friendly heating solution, and can utilize energy for heating from the ocean, a river, a well or the air. It can even be used to transfer the energy from one recirculation system to another, and thereby heat one system and cool another. Usually it utilizes energy from e.g. the ocean using a titanium heat exchanger, moves the energy to the recirculation that is calling for heating and releases the heat through another heat exchanger.

Different types of pumps are used for circulating the process water in the system. Pumping normally requires a substantial amount of electricity, and low lifting heights and efficient and correctly installed pumps are important to keep running costs at a minimum.

The lifting of water should preferably occur only once in the system, whereby the water runs by gravity all the way through the system back to the pump sump. Pumps are most often positioned in front of the biofilter system and the degasser as the water preparation process starts here. In any case, pumps should be placed after the mechanical filtration to avoid breaking the solids coming from the fish tanks.

Calculation of the total lifting height for pumping is the sum of the actual lifting height and the pressure losses in pipe runs, pipe bends and other fittings. This is also called the dynamic head. If water is pumped through a submerged biofilter before falling down through the degasser, a counter pressure from the biofilter will also have to be accounted for.

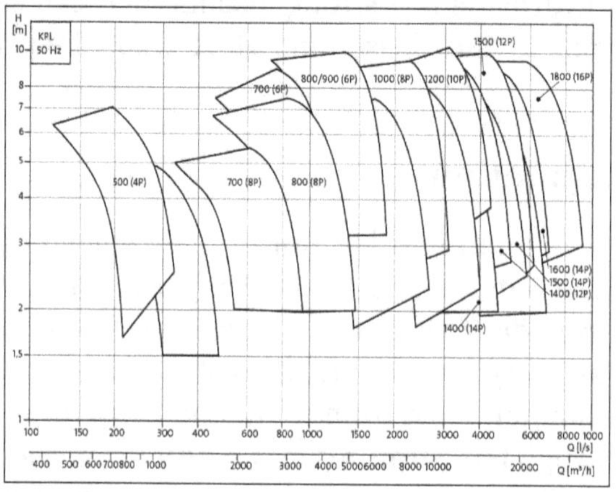

Figure: Lifting pumps type KPL for efficient lifting of large amounts of water.

In figure lifting pumps are often used for pumping the main flow in the recirculation system. Correct selection of pump is important to keep the running costs down. Frequency control is an option to regulate the exact flow needed depending on the fish production. H is the lifting height and Q is the volume of water lifted.

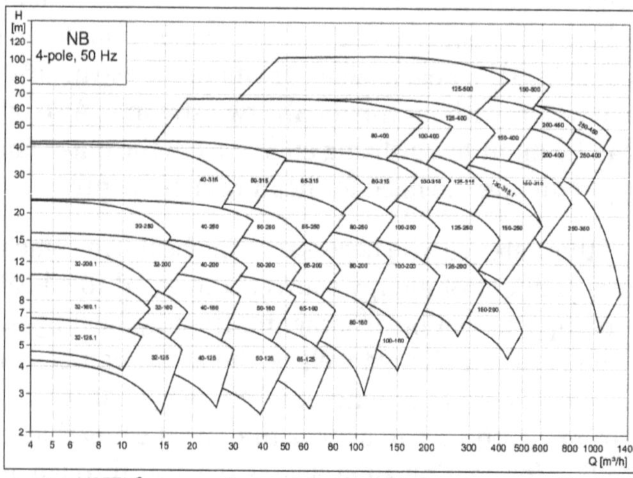

Figure: Centrifugal pumps type NB for pumping water when high pressure or high lifting heights are needed.

In figure the range of centrifugal pumps is wide, so these pumps are also efficiently used for pumping at lower lifting heights. Centrifugal pumps are often used in recirculation systems for pumping secondary flows as for example flows through UV systems or for reaching high pressure in oxygen cones. H is the lifting height and Q is the volume of water lifted.

The total lifting height in most intensive recirculation systems today is around 2-3 metres, which makes the use of low pressure pumps most efficient for pumping the main flow around. However, the process of dissolving pure oxygen into the process water requires centrifugal pumps as these pumps are able to create the required high pressure in the cone. In some systems, where the lifting height for the main flow is very low, the water is driven without the use of pumps by blowing air into aeration wells. In these systems the degassing and the movement of water are accomplished in one process, which makes low lifting heights possible. The efficiency of degassing and moving of water is however not necessarily better than that of pumping water up over the degasser, because the efficiency of aeration wells in terms of using energy and the degassing efficiency is lower than using lifting pumps and stripping or trickling the water.

Monitoring, Control and Alarms

Intensive fish farming requires close monitoring and control of the production in order to maintain optimal conditions for the fish at all times. Technical failures can easily result in substantial losses, and alarms are vital installations for securing the operation.

In many modern farms, a central control system can monitor and control oxygen levels, temperature, pH, water levels and motor functions. If any of the parameters moves out of the preset hysteresis values, a start/stop process will try to solve the problem. If the problem is not solved automatically, an alarm will start. Automatic feeding can also be an integrated part of the central control system. This allows the timing of the feeding to be coordinated precisely with a higher dosage of oxygen as the oxygen consumption rises during feeding. In less sophisticated systems, the monitoring and control is not fully automatic, and personnel will have to make several manual adjustments.

Figure: An oxygen probe (Oxyguard) is calibrated in the air before being lowered into the water for on-line measurement of the oxygen content of the water.

Whatever the case, no system will work without the surveillance of the personnel working on the farm. The control system must therefore be fitted with an alarm system, which will call the personnel if any major failures are about to occur. A reaction time of less than 20 minutes is recommended, even in situations where automatic back-up systems are installed.

Emergency System

The use of pure oxygen as a back-up is the number one safety precaution. The installation is simple, and consists of a holding tank for pure oxygen and a distribution system with diffusers fitted in all tanks. If the electricity supply fails a magnetic valve pulls back and pressurized oxygen flows to each tank keeping the fish alive. The flow sent to the diffusers should be adjusted beforehand, so that the oxygen in the storage tank in an emergency situation lasts long enough for the failure to be corrected in time.

Figure: Oxygen tank and emergency electrical generator.

To back up the electrical supply, a fuel driven electrical generator is necessary. It is very important to get the main pumps in operation as fast as possible, because ammonia excreted from the fish will build up to toxic levels when the water is not circulating over the biofilter. It is therefore important to get the water flow up and running within an hour or so.

Intake Water

Water used for recirculation should preferably come from a disease-free source or be sterilised before going into the system. In most cases it is better to use water from a borehole, a well, or something similar than to use water coming directly from a river, lake or the sea. If a treatment system for intake water needs to be installed, it will typically consist of a sand filter for microfiltration and a UV or ozone system for disinfection.

Fish Species in Recirculation

A recirculation system is a costly affair to build and to operate. There is competition on markets for fish and production must be efficient in order to make a profit. Selecting the right species to produce and constructing a well-functioning system are therefore of high importance. Essentially, the aim is to sell the fish at a high price and at the same time keep the production cost at the lowest possible level.

Water temperature is one of the most important parameters when looking at the feasibility of fish farming, because fish are cold blooded animals. This means that fish have the same body temperature as the temperature of the surrounding water. Fish cannot regulate their body temperature like pigs, cows or other farmed animals. Fish simply do not grow well when the water is cold; the warmer the water, the better the growth. Different species have different growth rates depending on the water temperature, and fish also have upper and lower lethal temperature limits. The farmer must be sure to keep his stock within these limits or the fish will die.

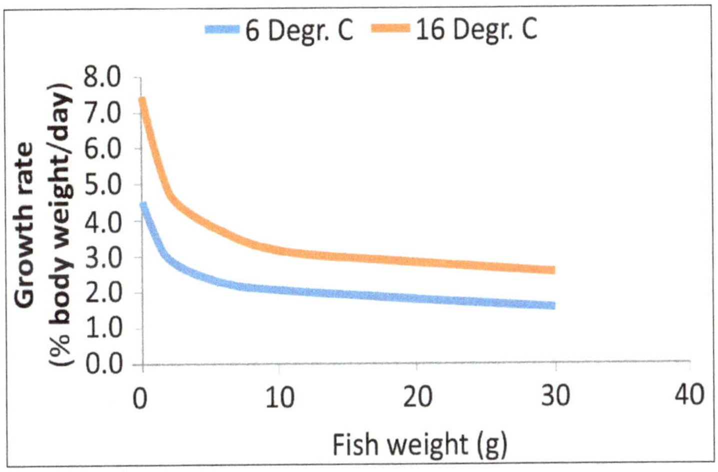

Figure: Growth rate of rainbow trout at 6 degrees and at 16 degrees Celsius as function of fish size.

Another issue affecting the feasibility of fish farming is the size of the fish grown in the farm. At any given temperature, small fish have a higher growth rate than large fish. This means that small fish are able to gain more weight over the same period of time than large fish.

Small fish also convert fish feed at a better rate than large fish - see figure. Growing faster and utilising feed more efficiently will of course have a positive influence on the production costs as these are lowered when calculated per kilo of fish produced. However, the production of small fish is just one step in the whole production process through to marketable fish. Naturally, not all fish produced in fish farming can be small fish, and the potential for growing small fish is therefore limited. Nevertheless, when discussing what kind of fish to produce in recirculation systems, the answer, first and foremost, will be small fish. It simply makes sense to invest money in fry production, because you get more out of your investment when farming small fish.

The cost of reaching and maintaining the optimal water temperature all year round in a recirculation facility is money well spent. Keeping fish at optimal rearing conditions will give a much higher growth rate in comparison to the often sub-optimal conditions in the wild. Also, it is important to note that all the advantages of clean water, sufficient oxygen levels, etc. in a recirculation system have a positive effect on survival rate, fish health, etc., which in the end gives a high quality product.

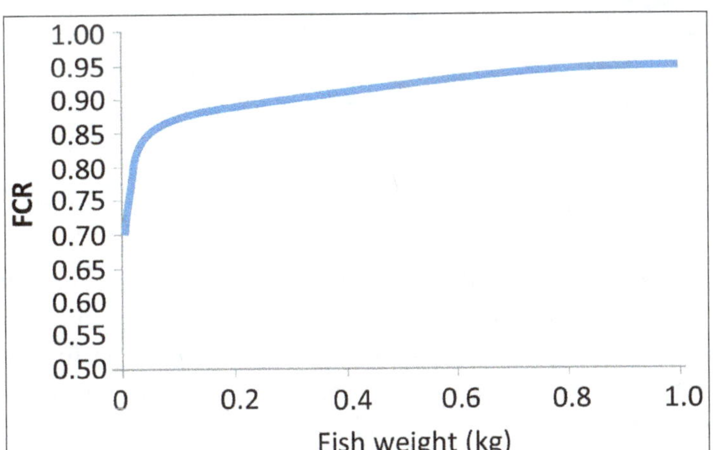

Figure: Feed conversion rate (FCR) of rainbow trout in a recirculation system, related to fish weight at 15-18 degrees Celsius.

Compared to other farmed animals there is a large variety of fish, and many different fish species are farmed. In comparison, the market for pigs, cattle or chicken is not diversified in the same way as fish. The consumer does not ask for different species of pigs, cattle or chicken, they just ask for different cuts or sizes of cuts. But when it comes to fish, the choice of species is wide, and the consumer is used to choosing from a range of different fish, a situation which makes many different fish species interesting in the eyes of any fish farmer. Over the past decade some hundred aquatic species have been introduced to aquaculture and the rate of domestication of aquatic species is around hundred times faster than that of the domestication of plants and animals on land.

Looking at the world production volume of farmed fish, the picture is not in favour of a multi species output. From figure below it can be seen that carp, of which we are only talking of some 5 different sub-species, is by far the most dominant. Salmon and trout are next in line, and this is only two species. The rest amounts to some ten species. One therefore has to realise that although there are plenty of species to be cultured, only a few of these go on to become real successes on a world-wide scale. However, this does not mean that all the new fish species introduced to aquaculture are failures. One just has to realise that the world production volume of new species is limited, and that the success and failures of growing these species depend very much on market conditions. Producing a small volume of a prestigious fish species may well be profitable as it fetches a high price. However, because the market for prestigious species is limited, the price may soon go down if production and thereby availability of the product rises. It can be very profitable to be the first and only one on the market with a new species in aquaculture. On the other hand, it is also a risky business with a high degree of uncertainty in both production and in market development.

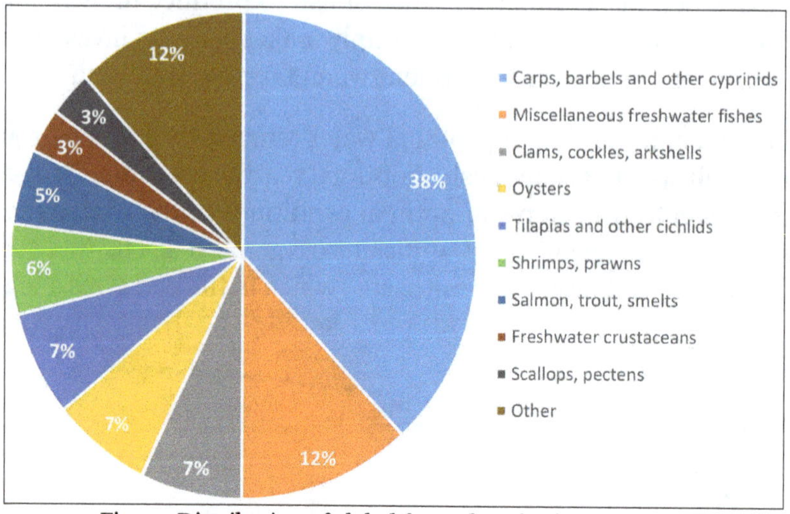

Figure: Distribution of global farmed seafood production.

When introducing new species in aquaculture it should also be remembered that it is wild species, which are being captured and tested in aquaculture. Domestication is most often a long and troublesome task. There are many impacts, which will influence growth performance, such as high genetic variation in growth rate, feed conversation rate, survival rate and problems with early maturation and disease susceptibility. Thus it is very likely that the performance of fish from the wild does not correspond to the expectations of the aquaculturist. Also, viruses in wild stocks can be brought in, of which some only appear after several years breeding, resulting in a demoralising experience.

To give general recommendations on which species to culture in recirculation systems is not an easy task. Many factors influence the success of a fish farming business. For example, local building costs, cost and stability of electricity supply, availability of skilled personnel, etc. Two important questions though should be asked before anything else is discussed: does the fish species being considered have the ability to perform well in a recirculation facility; and secondly is there a market for this species that will fetch a price high enough and at volumes large enough to make the project profitable.

The first question can be answered in a relatively simple manner: seen from a biological point of view, any type of fish reared successfully in traditional aquaculture can just as easily be reared in recirculation. As mentioned, the environment inside the recirculated fish farm can be adjusted to match the exact needs of the species reared. The recirculation technology in itself is not an obstacle to any new species introduced. The fish will grow just as well, and often even better, in a recirculation unit. Whether it will perform well from an economic point of view is more uncertain as this depends on the market conditions, the investment and the production costs and the ability of the species to grow rapidly. Rearing fish with generally low growth rates, such as extreme cold water species, makes it difficult to produce a yearly output that justifies the investment made in the facility.

Whether market conditions are favourable for a given species reared in a recirculation system depends highly on competition from other producers. And this is not restricted to local producers; fish trading is a global business and competition is global too. Trout farmed in Poland may well have to compete with catfish from Vietnam or salmon from farms in Norway as fish are easily distributed around the world at low cost.

It has always been recommended to use recirculation systems to produce expensive fish, because a high selling price leaves room for higher production costs. A good example is the eel farming business where a high selling price allows relatively high production costs. On the other hand, there is a strong tendency to use recirculation systems also for lower priced fish species such as trout or salmon.

The Danish recirculation trout farm concept is a good example of recirculation systems entering a relatively low price segment such as portion sized trout. However, it is necessary for such production systems to be huge, operating in volumes from 1 000 tonnes and upwards, in order to be competitive. In the future, perhaps in some areas growing large salmon will move from sea cage farming to land-based recirculation facilities for environmental reasons. Even an extremely low priced fish product such as tilapia will probably become profitable to grow in some kind of recirculation system as the fight for water and space intensifies.

The suitability of rearing specific fish species in recirculation depends on many different factors, such as the profitability, environmental concerns, biological suitability. In the tables below fish species have been grouped into different categories depending on the commercial feasibility of growing them in a recirculation system.

It should be mentioned that for small fish the use of recirculation is always recommended, because small fish grow faster and are therefore particularly suited to a controlled environment until they have reached the size for on- growing.

Good biological performance and acceptable market conditions make the following fish interesting for production to market size in recirculation aquaculture.

Species	Current status	Market
Arctic char (Salvelinus alpinus) 14°C 	Arctic char or cross breeds with brook trout has a long track record of growing well in cold water aquaculture.	Sold in specific markets at fair to good prices.
Atlantic salmon, smolt (Salmo salar) 14°C 	Small salmon are called smolt. They are grown in freshwater before transfer to saltwater for grow-out. Smolts are raised in recirculation systems with great success.	The market for salmon smolt is usually very good. Demand is constantly increasing.
Eel (Anguilla anguilla) 24°C 	Proven successful species in recirculation. Cannot reproduce in captivity. Wild catch of fry (elvers) is necessary. Considered threatened species.	Limited market with varying price levels. Some buyers will refuse to buy because of threatened species status.
Grouper (Epinephelus spp.) 28°C 	Saltwater fish grown primarily in Asia. Many different grouper species. Requires knowledge in spawning and larval rearing. Grow-out relatively straight forward.	Sold primarily in local markets at good prices in areas where production by many small producers is taking place.
Rainbow trout (Oncorhynchus mykiss) 16°C 	Easy to culture. Recirculation in freshwater widely used from fry rearing up to portion size fish. Larger trout can also be grown in recirculation whether fresh or saltwater.	Relatively tough competition in most markets. Products need to be diversified.
Seabass/ Seabream (Dicentrarchus labrax / Sparus aurata) 24°C 	Saltwater aquaculture fish in a highly developed cage farming industry. Larval phases require good skills. Proven to grow well in recirculation.	Generally tough market conditions, but can fetch good prices for fresh fish in some local areas.
Sturgeon (Acipenser spp.) 22°C 	Group of freshwater fish of many species relatively easy to culture. Skills required in different biological stages. Farming in recirculation systems is increasing.	Fair market conditions for meat. The caviar business seems to expand in high-end markets.

Species	Current status	Market
Turbot *(Scophthalmus maximus)* 17 °C 	Good skills required in broodstock and hatchery management. Grows very well in recirculation.	Generally tough international market conditions. Local market prices can be higher.
Whiteleg shrimp *(Penaeus vannamei)* 30 °C 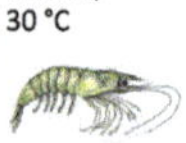	Most common shrimp species in aquaculture. Grow-out in recirculation systems has been proven successful. The production method is developing.	Shrimp prices are generally good and high in comparison to fish prices.
Yellowtail amberjack *(Seriola lalandi)* 22 °C 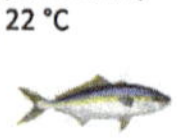	Yellowtail amberjack, or kingfish, is a saltwater species proven to perform well in cages and in land based systems.	Market prices good. Sold in specific markets.

Low market prices make the following fi challenging to produce with a profit in recirculate aquaculture, and good marketing and sales efforts are important:

Species	Current status	Market
African catfish *(Clarias gariepinus)* 28 °C 	Freshwater fish very easy to culture. A robust and fast growing fish that performs well in recirculation. Production must be very cost efficient.	Moderate to low prices. Most fish are sold live in local markets. Strong marketing effort required.
Barramundi *(Lates calcarifer)* 28 °C 	Also called Asian seabass. Lives in both fresh and saltwater. Requires knowledge in larval rearing. Relatively straight forward in grow-out.	Sold primarily in local markets at fair prices. International market expected to grow as global marketing increases.
Carps *(Cyprinus carpio)* 26 °C 	All carp species will grow very well in recirculation aquaculture systems. Keeping down production costs at a minimum is the main issue.	Carps are regarded as a low price species in most markets, but can fetch higher prices in some markets.
Pangasius *(Pangasius bocourti)* 28 °C 	This catfish is grown in big earth ponds primarily in Vietnam. Impressive ability to survive and grow at sub-optimal conditions.	Low end product in the global fish market leaves no room for production costs.
Perch *(Perca fluviatilis)* 17 °C 	Freshwater fish proven to grow well in recirculation although not widely used.	Limited market with fluctuating prices.
Tilapia *(Oreochromis niloticus)* 28 °C 	One of the most predominant aquaculture fish, which is robust and fast growing. Production cost must be kept to a minimum to be competitive.	Sold in the world market at low to moderate prices. Can fetch higher prices locally.
Whitefish *(Coregonus lavaretus)* 15 °C 	Coregonus is a group of freshwater fish that can be grown in aquaculture and in recirculation systems.	Prices relatively low as there is strong competition from wild caught species.

Very challenging to grow these fish at a commercial viable scale in recirculation aquaculture or in aquaculture in general, because it is either difficult to manage biologically or/and because of tough market conditions.

Species	Current status	Market
Atlantic cod *(Gadus morhua)* 12 °C 	Fry rearing proven to be successful in recirculation. Grow-out of larger cod needs further development and is as such not suited for recirculation.	Prices are fluctuating as market is heavily affected by wild stock catches.
Atlantic salmon, Large *(Salmo salar)* 14 °C 	Larger salmon are grown in sea cages to reach market size of 4-5 kilos. Grow-out in land based systems using recirculation is under development.	Global market dominated by Norwegian marketing. Trend towards certified products.
Bluefin tuna *(Thunnus thynnus)* 24 °C 	Fattening of wild caught fish is the only profitable farming technology. Controlling full cycle at a commercial level in aquaculture is still under development.	Can fetch very high prices in a turbulent worldwide market for tuna.
Cobia *(Rachycentron canadum)* 28 °C 	Fairly new saltwater aquaculture fish of good meat quality. Grow-out in cage culture. Output seems to be growing, although there are many obstacles in breeding.	Market is not well developed and the fish is unknown in most markets.
Lemon sole *(Microstomus kitt)* 17 °C 	Not yet fully developed new species in aquaculture due to different obstacles in biology, such as feeding, etc.	High-end product fetching stable and high prices.
Pike-perch *(Sander lucioperca)* 20 °C 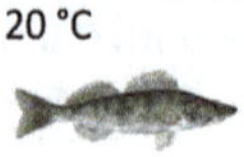	Freshwater fish difficult to farm. Larval stage troublesome, grow-out seems a little easier. Only a few successful recirculation systems for pike-perch.	Good and fair prices. Demand expected to grow as wild stocks fall and consumption increases.

Fish Slaughter

Slaughter is the term used when animals are killed for food. The methods used to slaughter fish depend on many factors, including fish species and size, fish numbers, type of aquaculture production system (e.g. ponds, cages, tanks), animal welfare concerns, regulatory compliance, economics, market preferences and effects on product quality. Some methods may not be approved in some countries, such as the use of certain anesthetics and other methods to kill fish.

Several organizations have prepared guidelines for humane treatment prior to and during slaughter, and slaughter methods may soon be specified in the purchase requirements of major seafood buyers. All methods of slaughter have issues where animal welfare concerns should be addressed. Approaches such as bleeding without stunning have been cited as less acceptable. Meanwhile, electrical stunning followed by an acceptable killing method, a combination that minimizes physical and biological effects, may have less animal welfare issues.

Selecting a Slaughter Method

Prior to slaughter, feed is withdrawn from cultured fish to allow their digestive systems to empty their contents. This reduces the chance of fish tissues being contaminated by gut contents during processing and helps to maintain the quality and hygiene of the final products.

Fish should not be crowded. Crowding exposes fish to a rapidly increasing density, and as a result, oxygen availability and water quality can decrease rapidly. Overly crowded fish may show vigorous agitation, gasping and splashing. There is also an increased risk of damage from abrasion through contact with other fish, nets or holding tanks. The adverse effects of crowding can be lessened by reducing densities slowly and providing additional oxygen. Fish should be crowded for as short a period of time as possible.

Harvesting fish for slaughter usually requires some handling and concentration of fish, which creates stress for the fish. Harvesting can cause elevated levels of cortisol, which is the primary stress hormone in fish, as well as lactic acid and glucose. Stress can also cause reduced glycogen levels, decreased muscle pH and rapid onset of rigor mortis. Thus, undesirable pre-harvest physiological changes that stress the fish can result in lower product quality and reduced processing yields. This can have a significant effect on profitability.

Slaughter Methods

A variety of slaughter methods are currently used for fish, depending on the species, desired product quality and market demand. Some fish are killed and processed individually, while others are killed and processed collectively. Slaughter is generally a twostage process. First, the fish are stunned so they cannot sense pain. Then the fish are killed by one of a variety of methods. It is important that the stun-to-kill time is brief so that fish do not regain consciousness before they are killed.

When choosing methods for slaughter, it is important to evaluate how each method will impact the fish, personnel, and resulting product. For each slaughter situation, be sure the method:

- Induces rapid loss of consciousness and death in fish.
- Creates a minimal amount of pain and distress.
- Produces irreversible loss of consciousness.
- Works with intended fish use and purpose.
- Provides reliable slaughter methodology.
- Works with species, size and age of fish.
- Is appropriate for the subsequent evaluation or use of fish tissues.
- Causes minimal environmental impact of fish carcasses be consumed.
- Is safe for personnel who use methodology.
- Minimizes emotional effect on observers or operators.

- Causes no damage to equipment.
- Meets regulatory requirements.

Fish Welfare and Common Slaughter Methods

Acceptable slaughter methods must render the animals insensible immediately and should be performed without causing undue pain or suffering. Rapid slaughter methods are preferred over slower methods, since they lead to improved fish welfare and fish quality.

Fish Slaughter Methods

Carbon Dioxide

Carbon dioxide saturated water causes narcosis and loss of consciousness in the fish. Only carbon dioxide from a source that allows for careful regulation of the concentration, such as from a gas cylinder, is acceptable.

The gas is dissolved in the water prior to fish being introduced to the water. Fish placed in this water often react violently while their blood absorbs the gas rapidly. The fish may get bruised from hitting other fish or the sides of the holding container. The time required for fish to be killed can vary from a few minutes to 10+ minutes. Thus, carbon dioxide is often used as the first step in a two-step process. First, fish are anesthetized by the carbon dioxide. Then, fish are killed rapidly by a secondary method. Some countries do not allow the use of carbon dioxide due to welfare concerns.

Captive Bolt

With the captive bolt method, fish are oriented into a machine individually that uses a pneumatic non-penetrating captive bolt to stun the fish. This technique is accurate and delivers the required concussion velocity to stun the fish quickly. Fish are then killed manually or mechanically by exsanguination using an acceptable bleeding technique, such as cutting or removing the gills.

Gunshot

This ballistic technique is used with large wild-caught fish species such as halibut and tuna. The bullet penetrates the brain, where it causes extensive damage and immediate death of the fish. Inaccuracy and human safety are major concerns of this technique.

Spiking

In spiking, pithing, coring or "ikejime", a spike or probe is inserted through the skull of the fish directly into the brain. The fish are stunned and killed at the same time. This procedure can be applied more accurately in large fish due to the larger size of the brain. In smaller fish, the brain may be difficult to locate and destroy. If the brain is not adequately destroyed, the fish can undergo distress, and undesirable tissue quality changes may result.

Other Manual Techniques

These methods include manual blunt force trauma, decapitation and cervical transection. With blunt force trauma, fish are removed from the water individually and given a sharp blow to the head. If the blow is accurate and strong enough, the animal is killed instantly. However, if the blow is weaker, the animal is only stunned and a second step of killing is required. Both decapitation and transection of the cervical spine also require a secondary kill method such as pithing.

Electrical Stunning

Stunning by use of electricity is known as electronarcosis. Killing by electricity is known as electrocution. Electrocution completely destroys brain function and renders the animal unconscious while stopping the breathing reflex, so the fish is stunned and killed simultaneously. Electric stunning is reversible, as normal brain function is disrupted for only short period of time. Hence, electronarcosis must be followed immediately by a secondary kill method. Electric stunning immobilizes the fish and prevents distress and struggling prior to slaughter, which are detrimental to tissue quality. Recent advances in electrical equipment design have made substantial improvements in preventing or minimizing undesirable physical and biological effects in electric stunned fish. For electrical stunning to be effective, proper current and stun duration must be maintained. Also, water factors such as conductivity and temperature must be properly managed.

Bleeding

Killing a fish by bleeding can be accomplished by any of three major processes: cutting the gills, removing the gills or severing the caudal vessels of the tail. Fish should be stunned prior to bleeding. Fish die of anoxia and proper stunning should prevent any distress or struggling of the fish. Bleeding prevents fish muscles from turning an undesirable red color or acquiring a blood odor, which can prevent fish from being sold for sushi or surimi.

Rapid Chilling

Placing fish in iced water can be used to immobilize fish so they can be more easily handled for a secondary killing method. Fish can also be placed in an ice bath for a longer time until death occurs, but this technique is only appropriate for warm water fish species and not recommended for temperate, cool or cold water species, or for medium to large-bodied fish where the time to death may be significantly prolonged. Even though fish usually struggle prior to loss of consciousness, live chilling is considered by many in the aquaculture industry to offer benefits to carcass quality, because reducing muscle temperature close to zero degrees Celsius helps delay enzymatic and microbial spoilage processes. The process also increases the time for onset of rigor mortis and the resolution of rigor. Another advantage of this slaughter method is that the water can be drained and the fish placed in an iced container with the fish's temperature already lowered.

Asphyxiation

Asphyxiation by removal from water, or "dewatering", has been commonly used by the aquaculture industry for many years. Basically, fish are removed from the water and allowed to die via asphyxiation. The fish struggles as oxygen is depleted and respiration ceases. The rate at which oxygen depletion occurs a depends upon ambient temperature and the rate of fish activity (e.g. reducing the temperature of fish typically prolongs the time to anoxia and, therefore, the time to insensibility, thereby lengthening the period of distress). This method is less acceptable compared to the aforementioned methods and is not considered acceptable by American Veterinary Medical Association.

Anesthesia

With the exception of carbon dioxide, the use of anesthetics is not an acceptable method for slaughter sas there are regulatory and human consumption concerns over potential anesthetic residues in the fish tissues.

References

- "Fish Farms Drive Wild Salmon Populations Toward Extinction". ScienceDaily. 16 December 2007. Retrieved 2018-01-06
- Integrated-recycling-systems: fishnetafrica.org, Retrieved 30 March 2020
- Ford, JS; Myers, RA (2008). "A Global Assessment of Salmon Aquaculture Impacts on Wild Salmonids". PLOS Biol. 6 (2): e33. doi:10.1371/journal.pbio.0060033. PMC 2235905. PMID 18271629
- Fisheries-types-of-fisheries-and-it-economical-importance-1361: biologydiscussion.com, Retrieved 24 July 2020
- Huntingford, F. A; Adams, C; Braithwaite, V. A; Kadri, S; Pottinger, T. G; Sandoe, P; Turnbull, J. F (2006). "Current issues in fish welfare" (PDF). Journal of Fish Biology. 68(2): 332–372. doi:10.1111/j.0022-1112.2006.001046.x
- Fish-farming-105599: wellnessmama.com, Retrieved 14 April 2020
- Barrionuevo, Alexei (July 26, 2009). "Chile's Antibiotics Use on Salmon Farms Dwarfs That of a Top Rival's". The New York Times. Retrieved 2009-08-28

4

Diseases and Parasites Related to Fishes

Like any other commercial activity, aquaculture has its own challenges. The complications brought about by unwanted parasites and diseases associated with fishes can have devastating effects on aquaculture. This chapter discusses the different diseases and parasites which can harm fishes, with special attention being given to the salmon.

Fish Diseases and Parasites

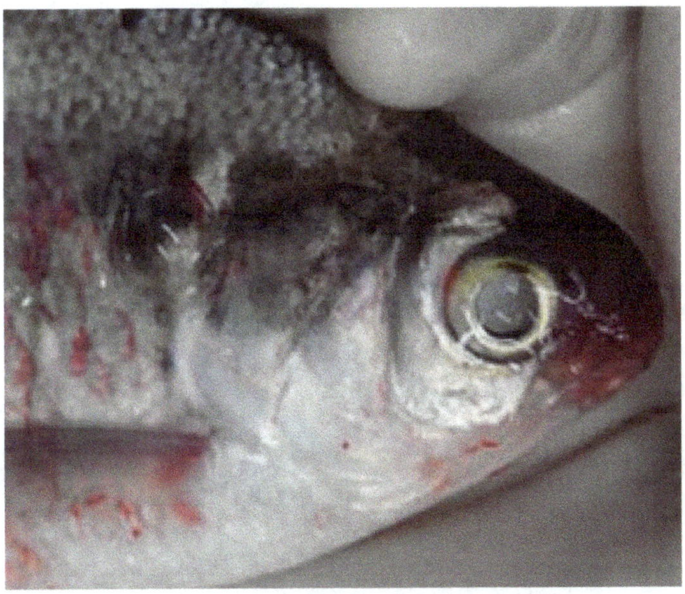

This gizzard shad has VHS, a deadly infectious disease which causes bleeding. It afflicts over 50 species of freshwater and marine fish in the northern hemisphere.

Like humans and other animals, fish suffer from diseases and parasites. Fish defences against disease are specific and non-specific. Non-specific defences include skin and scales, as well as the mucus layer secreted by the epidermis that traps microorganisms and inhibits their growth. If pathogens breach these defences, fish can develop inflammatory responses that increase the flow of blood to infected areas and deliver white blood cells that attempt to destroy the pathogens.

Specific defences are specialised responses to particular pathogens recognised by the fish's body, that is adaptative immune responses. In recent years, vaccines have become widely used in aquaculture and ornamental fish, for example vaccines for furunculosis in farmed salmon and koi herpes virus in koi.

This flatfish *Limanda limanda* has an outgrowth called a xenoma. It is caused by a microsporidian fungal parasite in its intestines.

Some commercially important fish diseases are VHS, ich and whirling disease.

Disease

A veterinarian gives an injection to a goldfish.

All fish carry pathogens and parasites. Usually this is at some cost to the fish. If the cost is sufficiently high, then the impacts can be characterised as a disease. However disease in fish is not understood well. What is known about fish disease often relates to aquaria fish, and more recently, to farmed fish.

Disease is a prime agent affecting fish mortality, especially when fish are young. Fish can limit the impacts of pathogens and parasites with behavioural or biochemical means, and such fish have reproductive advantages. Interacting factors result in low grade infection becoming fatal diseases. In particular, things that causes stress, such as natural droughts or pollution or predators, can precipitate outbreak of disease.

Disease can also be particularly problematic when pathogens and parasites carried by introduced species affect native species. An introduced species may find invading easier if potential predators and competitors have been decimated by disease.

Pathogens which can cause fish diseases comprise:

- Viral infections.
- Bacterial infections, such as *Pseudomonas fluorescens* leading to fin rot and fish dropsy.
- Fungal infections.
- Water mould infections, such as *Saprolegnia* sp.
- Metazoan parasites, such as copepods.
- Unicellular parasites, such as *Ichthyophthirius multifiliis* leading to ich.
- Certain parasites like Helminths for example *Eustrongylides*.

Parasites

The isopod Anilocra gigantea parasitising the snapper *Pristipomoides filamentosus*.

Parasites in fish are a common natural occurrence. Parasites can provide information about host population ecology. In fisheries biology, for example, parasite communities can be used to distinguish distinct populations of the same fish species co-inhabiting a region. Additionally, parasites possess a variety of specialized traits and life-history strategies that enable them to colonize hosts. Understanding these aspects of parasite ecology, of interest in their own right, can illuminate parasite-avoidance strategies employed by hosts.

Usually parasites (and pathogens) need to avoid killing their hosts, since extinct hosts can mean extinct parasites. Evolutionary constraints may operate so parasites avoid killing their hosts, or the natural variability in host defensive strategies may suffice to keep host populations viable. Parasite infections can impair the courtship dance of male threespine sticklebacks. When that happens, the females reject them, suggesting a strong mechanism for the selection of parasite resistance."

However, not all parasites want to keep their hosts alive, and there are parasites with multistage life cycles who go to some trouble to kill their host. For example, some tapeworms make some fish behave in such a way that a predatory bird can catch it. The predatory bird is the next host for the parasite in the next stage of its life cycle. Specifically, the tapeworm *Schistocephalus solidus* turns infected threespine stickleback white, and then makes them more buoyant so that they splash along at the surface of the water, becoming easy to see and easy to catch for a passing bird.

Cymothoa exigua is a parasitic crustacean which enters a fish through its gills and destroys the fish's tongue.

Parasites can be internal (endoparasites) or external (ectoparasites). Some internal fish parasites are spectacular, such as the philometrid nematode *Philometra fasciati* which is parasitic in the ovary of female Blacktip grouper; the adult female parasite is a red worm which can reach up to 40 centimetres in length, for a diameter of only 1.6 millimetre; the males are tiny. Other internal parasites are found living inside fish gills, include encysted adult didymozoid trematodes, a few trichosomoidid nematodes of the genus *Huffmanela*, including *Huffmanela ossicola* which lives within the gill bone, and the encysted parasitic turbellarian *Paravortex*. Various protists and Myxosporea are also parasitic on gills, where they form cysts.

Fish gills are also the preferred habitat of many external parasites, attached to the gill but living out of it. The most common are monogeneans and certain groups of parasitic copepods, which can be extremely numerous. Other external parasites found on gills are leeches and, in seawater, larvae of gnathiid isopods. Isopod fish parasites are mostly external and feed on blood. The larvae of the Gnathiidae family and adult cymothoidids have piercing and sucking mouthparts and clawed limbs adapted for clinging onto their hosts. *Cymothoa exigua* is a parasite of various marine fish. It causes the tongue of the fish to atrophy and takes its place in what is believed to be the first instance discovered of a parasite functionally replacing a host structure in animals.

Other parasitic disorders, include *Gyrodactylus salaris*, *Ichthyophthirius multifiliis*, cryptocaryon, velvet disease, *Brooklynella hostilis*, Hole in the head, *Glugea*, *Ceratomyxa shasta*, *Kudoa thyrsites*, *Tetracapsuloides bryosalmonae*, *Cymothoa exigua*, leeches, nematode, flukes, carp lice and salmon lice.

Although parasites are generally considered to be harmful, the eradication of all parasites would not necessarily be beneficial. Parasites account for as much as or more than half of life's diversity; they perform an important ecological role (by weakening prey) that ecosystems would take some time to adapt to; and without parasites organisms may eventually tend to asexual reproduction, diminishing the diversity of sexually dimorphic traits. Parasites provide an opportunity for the transfer of genetic material between species. On rare, but significant, occasions this may facilitate evolutionary changes that would not otherwise occur, or that would otherwise take even longer.

Below are some life cycles of fish parasites:

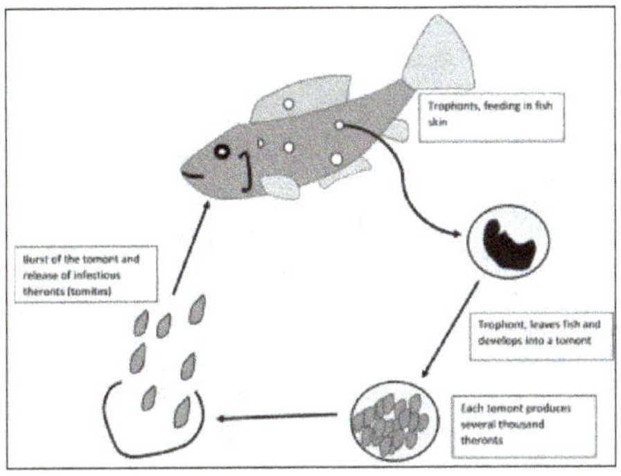

Life cycle of the fish parasite *Ichthyophthirius multifiliis*, commonly called ich.

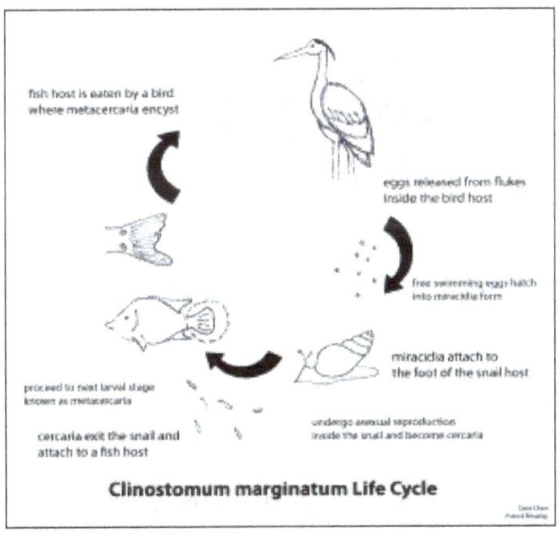

Life cycle of the parasitic fluke *Clinostomum marginatum*, commonly called the yellow grub.

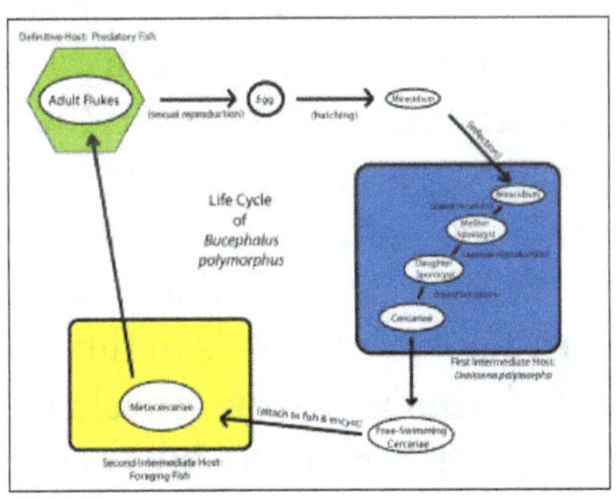

Life cycle of the digenean *Bucephalus polymorphus*.

Cleaner Fish

Some fish take advantage of cleaner fish for the removal of external parasites. The best known of these are the Bluestreak cleaner wrasses of the genus *Labroides* found on coral reefs in the Indian Ocean and Pacific Ocean. These small fish maintain so-called "cleaning stations" where other fish, known as hosts, will congregate and perform specific movements to attract the attention of the cleaner fish. Cleaning behaviours have been observed in a number of other fish groups, including an interesting case between two cichlids of the same genus, *Etroplus maculatus*, the cleaner fish, and the much larger *Etroplus suratensis*, the host.

Two cleaner wrasses, *Labroides phthirophagus*, servicing a goatfish, *Mulloidichthys flavolineatus*.

More than 40 species of parasites may reside on the skin and internally of the ocean sunfish, motivating the fish to seek relief in a number of ways. In temperate regions, drifting kelp fields harbour cleaner wrasses and other fish which remove parasites from the skin of visiting sunfish. In the tropics, the *mola* will solicit cleaner help from reef fishes. By basking on its side at the surface, the sunfish also allows seabirds to feed on parasites from their skin. Sunfish have been reported to breach more than ten feet above the surface, possibly as another effort to dislodge parasites on the body.

Mass Die Offs

Some diseases result in mass die offs. One of the more bizarre and recently discovered diseases produces huge fish kills in shallow marine waters. It is caused by the ambush predator dinoflagellate *Pfiesteria piscicida*. When large numbers of fish, like shoaling forage fish, are in confined situations such as shallow bays, the excretions from the fish encourage this dinoflagellate, which is not normally toxic, to produce free-swimming zoospores. If the fish remain in the area, continuing to provide nourishment, then the zoospores start secreting a neurotoxin. This toxin results in the fish developing bleeding lesions, and their skin flakes off in the water. The dinoflagellates then eat the blood and flakes of tissue while the affected fish die. Fish kills by this dinoflagellate are common, and they may also have been responsible for kills in the past which were thought to have had other causes. Kills like these can be viewed as natural mechanisms for regulating the population of exceptionally abundant fish. The rate at which the kills occur increases as organically polluted land runoff increases.

Wild Salmon

According to Canadian biologist Dorothy Kieser, protozoan parasite *Henneguya salminicola* is commonly found in the flesh of salmonids. It has been recorded in the field samples of salmon returning to the Queen Charlotte Islands. The fish responds by walling off the parasitic infection into a number of cysts that contain milky fluid. This fluid is an accumulation of a large number of parasites.

Henneguya salminicola, a parasite commonly found in the flesh of salmonids on the West Coast of Canada.

Henneguya and other parasites in the myxosporean group have a complex lifecycle where the salmon is one of two hosts. The fish releases the spores after spawning. In the *Henneguya* case, the spores enter a second host, most likely an invertebrate, in the spawning stream. When juvenile salmon out-migrate to the Pacific Ocean, the second host releases a stage infective to salmon. The parasite is then carried in the salmon until the next spawning cycle. The myxosporean parasite that causes whirling disease in trout, has a similar lifecycle. However, as opposed to whirling disease, the *Henneguya* infestation does not appear to cause disease in the host salmon – even heavily infected fish tend to return to spawn successfully.

According to Dr. Kieser, a lot of work on *Henneguya salminicola* was done by scientists at the Pacific Biological Station in Nanaimo in the mid-1980s, in particular, an overview report which states that "the fish that have the longest fresh water residence time as juveniles have the most noticeable infections. Hence in order of prevalence coho are most infected followed by sockeye, chinook, chum and pink." As well, the report says that, at the time the studies were conducted, stocks from the middle and upper reaches of large river systems in British Columbia such as Fraser, Skeena, Nass and from mainland coastal streams in the southern half of B.C. "are more likely to have a low prevalence of infection." The report also states "It should be stressed that *Henneguya*, economically deleterious though it is, is harmless from the view of public health. It is strictly a fish parasite that cannot live in or affect warm blooded animals, including man".

According to Klaus Schallie, Molluscan Shellfish Program Specialist with the Canadian Food Inspection Agency, "*Henneguya salminicola* is found in southern B.C. also and in all species of salmon. I have previously examined smoked chum salmon sides that were riddled with cysts and some sockeye runs in Barkley Sound (southern B.C., west coast of Vancouver Island) are noted for their high incidence of infestation."

Sea lice, particularly *Lepeophtheirus salmonis* and a variety of *Caligus* species, including *Caligus clemensi* and *Caligus rogercresseyi*, can cause deadly infestations of both farm-grown and wild salmon. Sea lice are ectoparasites which feed on mucous, blood, and skin, and migrate and latch onto the skin of wild salmon during free-swimming, planktonic *naupli* and *copepodid* larval stages, which can persist for several days. Large numbers of highly populated, open-net salmon farms can create exceptionally large concentrations of sea lice; when exposed in river estuaries containing large numbers of open-net farms, many young wild salmon are infected, and do not survive as a result. Adult salmon may survive otherwise critical numbers of sea lice, but small, thin-skinned juvenile salmon migrating to sea are highly vulnerable. On the Pacific coast of Canada, the louse-induced mortality of pink salmon in some regions is commonly over 80%.

Sample of pink salmon infected with Henneguya salminicola, caught off the Queen Charlotte Islands, Western Canada in 2009.

Farmed Salmon

In 1972, Gyrodactylus salaris, also called salmon fluke, a monogenean parasite, spread from Norwegian hatcheries to wild salmon, and devastated some wild salmon populations.

Atlantic salmon.

In 1984, infectious salmon anemia (ISAv) was discovered in Norway in an Atlantic salmon hatchery. Eighty percent of the fish in the outbreak died. ISAv, a viral disease, is now a major threat to the viability of Atlantic salmon farming. It is now the first of the diseases classified on List One of the European Commission's fish health regime. Amongst other measures, this requires the total eradication of the entire fish stock should an outbreak of the disease be confirmed on any farm. ISAv seriously affects salmon farms in Chile, Norway, Scotland and Canada, causing major economic losses to infected farms. As the name implies, it causes severe anemia of infected fish. Unlike mammals, the red blood cells of fish have DNA, and can become infected with viruses. The fish develop pale gills, and may swim close to the water surface, gulping for air. However, the disease can also develop without the fish showing any external signs of illness, the fish maintain a normal appetite, and then they suddenly die. The disease can progress slowly throughout an infected farm and, in the worst cases, death rates may approach 100 percent. It is also a threat to the dwindling stocks of wild salmon. Management strategies include developing a vaccine and improving genetic resistance to the disease.

In the wild, diseases and parasites are normally at low levels, and kept in check by natural predation on weakened individuals. In crowded net pens they can become epidemics. Diseases and parasites also transfer from farmed to wild salmon populations. A recent study in British Columbia links the spread of parasitic sea lice from river salmon farms to wild pink salmon in the same river." The European Commission (2002) concluded "The reduction of wild salmonid abundance is also linked to other factors but there is more and more scientific evidence establishing a direct link between the number of lice-infested wild fish and the presence of cages in the same estuary." It is reported that wild salmon on the west coast of Canada are being driven to extinction by sea lice from nearby salmon farms. Antibiotics and pesticides are often used to control the diseases and parasites.

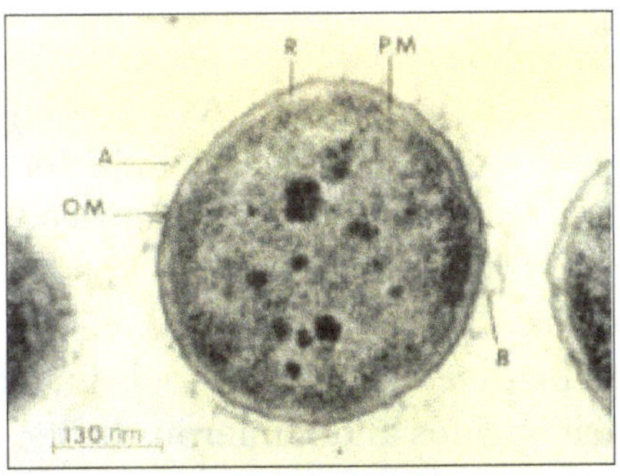

Aeromonas salmonicida, a Gram-negative bacteria, causes the disease furunculosis in marine and freshwater fish.

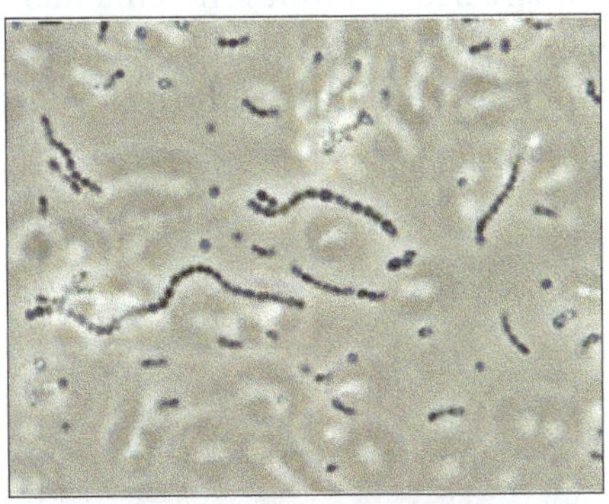

Streptococcus iniae, a Gram-positive, sphere-shaped bacteria caused losses in farmed marine and freshwater finfish of US$100 million in 1997.

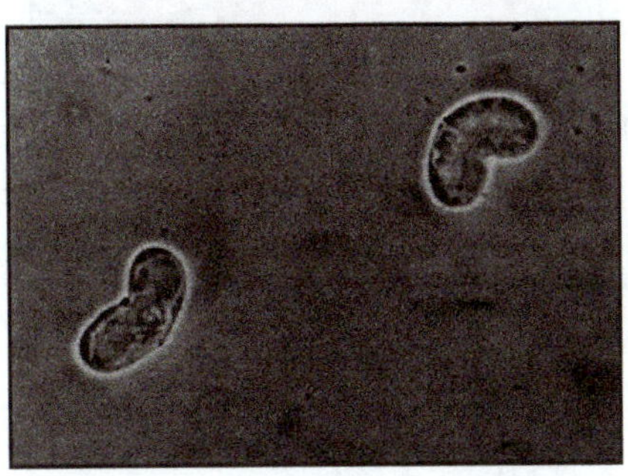

Ceratomyxa shasta, another myxosporean parasite, infects salmonid fish on the Pacific coast of North America.

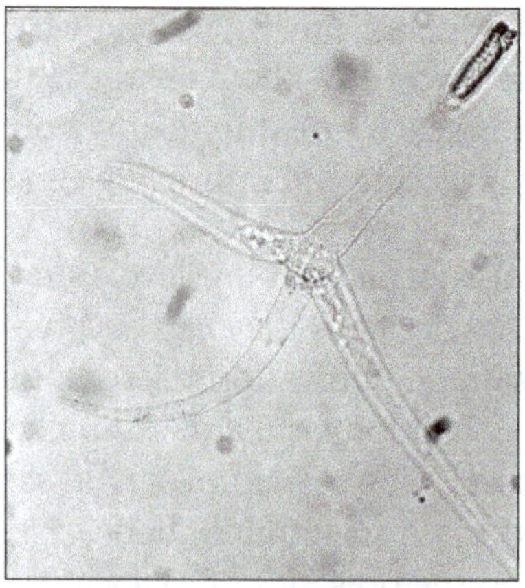

Myxobolus cerebralis, a myxosporean parasite, causes *whirling disease* in farmed salmon and trout and also in wild fish populations.

Coral Reef Fish

Coral reef fish are characterized by high biodiversity. As a consequence parasites of coral reef fish show tremendous variety. Parasites of coral reef fish include nematodes, Platyhelminthes (cestodes, digeneans, and monogeneans), leeches, parasitic crustaceans such as isopods and copepods, and various microorganisms such as myxosporidia and microsporidia. Some of these fish parasites have heteroxenous life cycles (i.e. they have several hosts) among which sharks (certain cestodes) or molluscs (digeneans). The high biodiversity of coral reefs increases the complexity of the interactions between parasites and their various and numerous hosts. Numerical estimates of parasite biodiversity have shown that certain coral fish species have up to 30 species of parasites. The mean number of parasites per fish species is about ten. This has a consequence in term of co-extinction. Results obtained for the coral reef fish of New Caledonia suggest that extinction of a coral reef fish species of average size would eventually result in the co-extinction of at least ten species of parasites.

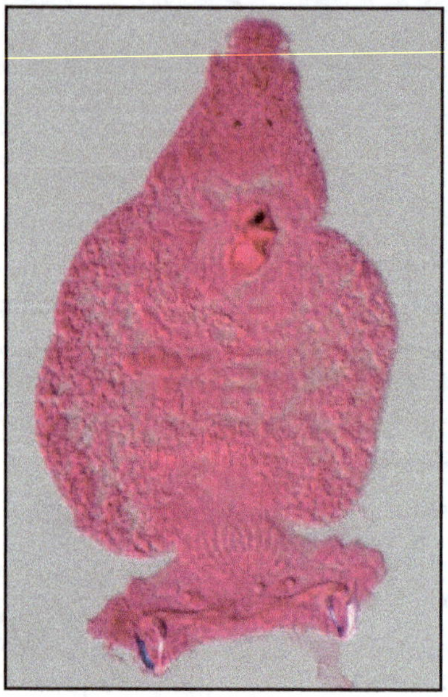

Monogenean parasite on the gill of a grouper.

Aquarium Fish

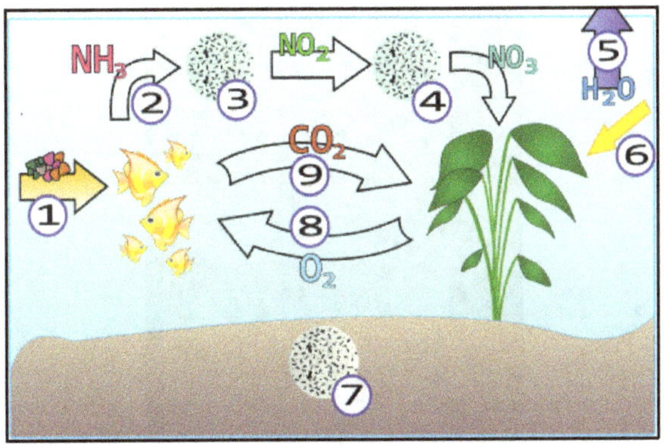

Nitrogen cycle in a common aquarium.

In most aquarium tanks, the fish are at high concentrations and the volume of water is limited. This means that communicable diseases can spread rapidly to most or all fish in a tank. An improper nitrogen cycle, inappropriate aquarium plants and potentially harmful freshwater invertebrates can directly harm or add to the stresses on ornamental fish in a tank. Despite this, many diseases in captive fish can be avoided or prevented through proper water conditions and a well-adjusted ecosystem within the tank. Ammonia poisoning is a common disease in new aquariums, especially when immediately stocked to full capacity.

Ornamental fish kept in aquariums are susceptible to numerous diseases.

Due to their generally small size and the low cost of replacing diseased or dead aquarium fish, the cost of testing and treating diseases is often seen as more trouble than the value of the fish.

Goldfish with dropsy.

Columnaris in the gill of a chinook salmon.

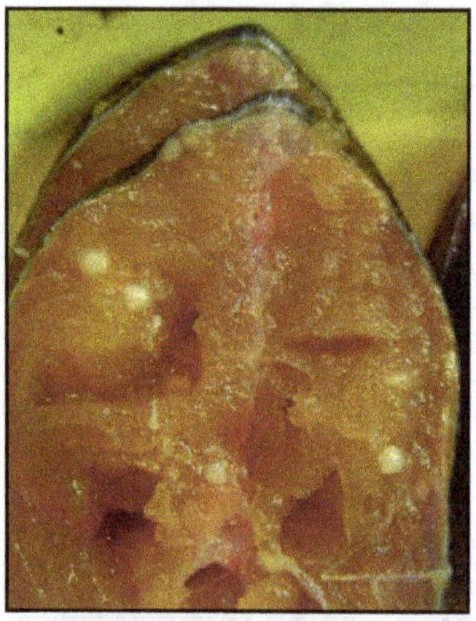

The parasite *Henneguya zschokkei* in salmon beard.

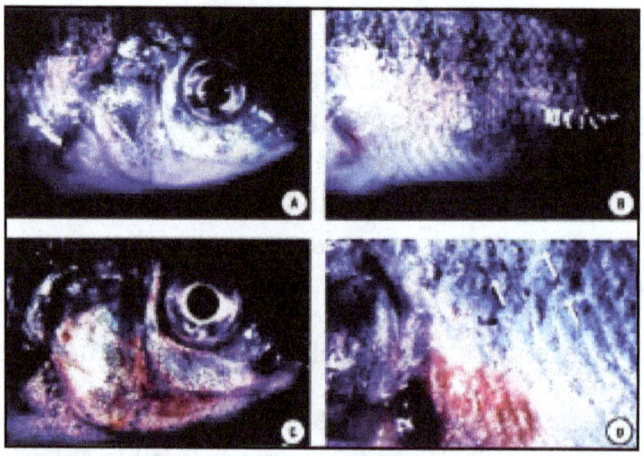

Skin ulcers in tilapia exposed to Pfiesteria shumwayae.

Immune System

Immune organs vary by type of fish. In the jawless fish (lampreys and hagfish), true lymphoid organs are absent. These fish rely on regions of lymphoid tissue within other organs to produce immune cells. For example, erythrocytes, macrophages and plasma cells are produced in the anterior kidney (or pronephros) and some areas of the gut (where granulocytes mature.) They resemble primitive bone marrow in hagfish. Cartilaginous fish (sharks and rays) have a more advanced immune system. They have three specialized organs that are unique to chondrichthyes; the epigonal organs (lymphoid tissue similar to mammalian bone) that surround the gonads, the Leydig's organ within the walls of their esophagus, and a spiral valve in their intestine. These organs house typical immune cells (granulocytes, lymphocytes and plasma cells). They also possess an identifiable thymus and a well-developed spleen (their most important immune organ) where various lymphocytes, plasma cells and macrophages develop and are stored. Chondrostean fish (sturgeons, paddlefish and bichirs) possess a major site for the production of granulocytes within a mass that is associated with the meninges (membranes surrounding the central nervous system.) Their heart is frequently covered with tissue that contains lymphocytes, reticular cells and a small number of macrophages. The chondrostean kidney is an important hemopoietic organ; where erythrocytes, granulocytes, lymphocytes and macrophages develop.

Like chondrostean fish, the major immune tissues of bony fish (or teleostei) include the kidney (especially the anterior kidney), which houses many different immune cells. In addition, teleost fish possess a thymus, spleen and scattered immune areas within mucosal tissues (e.g. in the skin, gills, gut and gonads). Much like the mammalian immune system, teleost erythrocytes, neutrophils and granulocytes are believed to reside in the spleen whereas lymphocytes are the major cell type found in the thymus. In 2006, a lymphatic system similar to that in mammals was described in one species of teleost fish, the zebrafish. Although not confirmed as yet, this system presumably will be where naive (unstimulated) T cells accumulate while waiting to encounter an antigen.

Spreading Disease and Parasites

The capture, transportation and culture of bait fish can spread damaging organisms between ecosystems, endangering them. In 2007, several American states, including Michigan, enacted regulations designed to slow the spread of fish diseases, including viral hemorrhagic septicemia, by bait fish. Because of the risk of transmitting *Myxobolus cerebralis* (whirling disease), trout and salmon should not be used as bait. Anglers may increase the possibility of contamination by emptying bait buckets into fishing venues and collecting or using bait improperly. The transportation of fish from one location to another can break the law and cause the introduction of fish and parasites alien to the ecosystem.

Eating Raw Fish

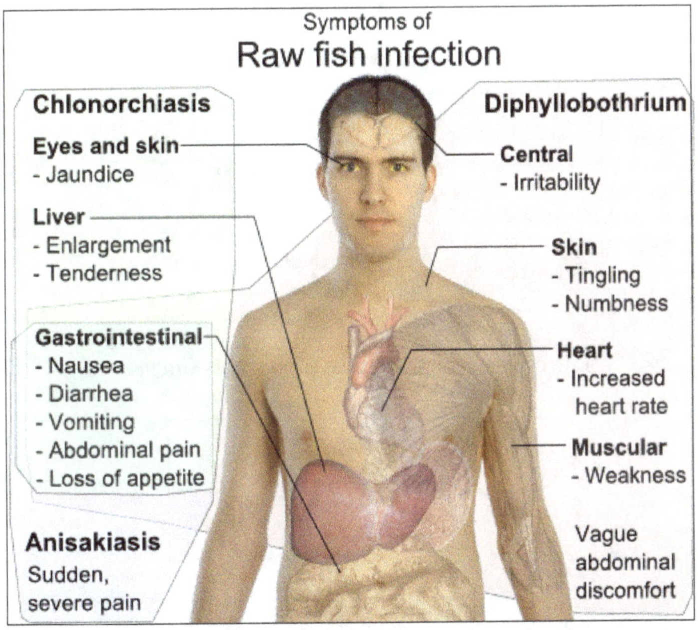

Differential symptoms of parasite infection by raw fish: Clonorchis sinensis (a trematode/fluke), Anisakis (a nematode/roundworm) and Diphyllobothrium a (cestode/tapeworm), all have gastrointestinal, but otherwise distinct, symptoms.

Though not a health concern in thoroughly cooked fish, parasites are a concern when human consumers eat raw or lightly preserved fish such as sashimi, sushi, ceviche, and gravlax. The popularity of such raw fish dishes makes it important for consumers to be aware of this risk. Raw fish should be frozen to an internal temperature of −20 °C (−4 °F) for at least 7 days to kill parasites. It is important to be aware that home freezers may not be cold enough to kill parasites.

Traditionally, fish that live all or part of their lives in fresh water were considered unsuitable for sashimi due to the possibility of parasites. Parasitic infections from fresh-water fish are a serious problem in some parts of the world, particularly Southeast Asia. Fish that spend part of their life cycle in salt water, like salmon, can also be a problem. A study in Seattle, Washington showed that 100% of wild salmon had roundworm larvae capable of infecting people. In the same study farm raised salmon did not have any roundworm larvae.

Parasite infection by raw fish is rare in the developed world (fewer than 40 cases per year in the U.S.), and involves mainly three kinds of parasites: Clonorchis sinensis (a trematode/fluke), Anisakis (a nematode/roundworm) and Diphyllobothrium (a cestode/tapeworm). Infection by the fish tapeworm *Diphyllobothrium latum* is seen in countries where people eat raw or undercooked fish, such as some countries in Asia, Eastern Europe, Scandinavia, Africa, and North and South America. Infection risk of anisakis is particularly higher in fishes which may live in a river such as salmon (*shake*) in Salmonidae, mackerel (*saba*). Such parasite infections can generally be avoided by boiling, burning, preserving in salt or vinegar, or freezing overnight. Even Japanese people never eat raw salmon or ikura (salmon roe), and even if they seem raw, these foods are not raw but are frozen overnight to prevent infections from parasites, particularly anisakis.

Below are some life cycles of fish parasites that can infect humans:

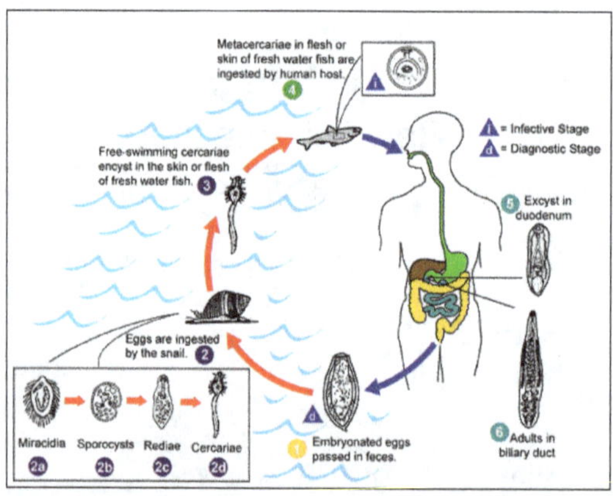

Life cycle of the liver fluke *Clonorchis sinensis.*

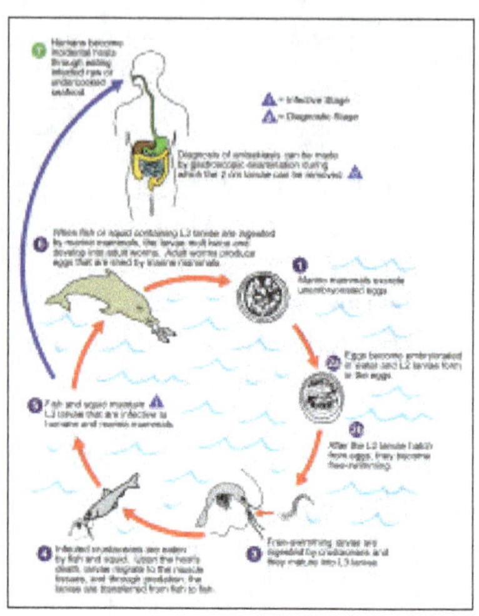

Life cycle of the parasitic *Anisakis* worm.

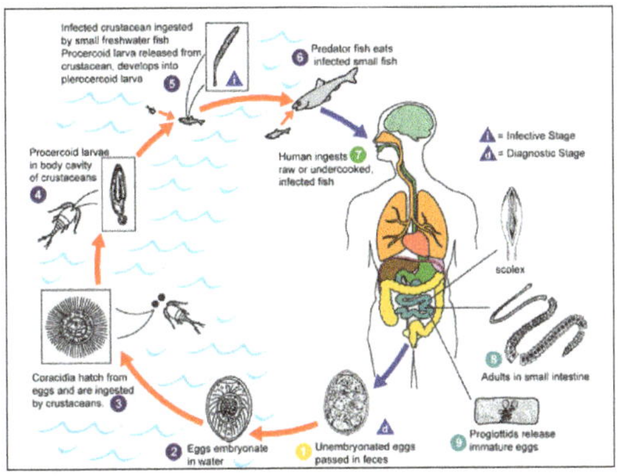

Life cycle of the fish tapeworm *Diphyllobothrium latum*.

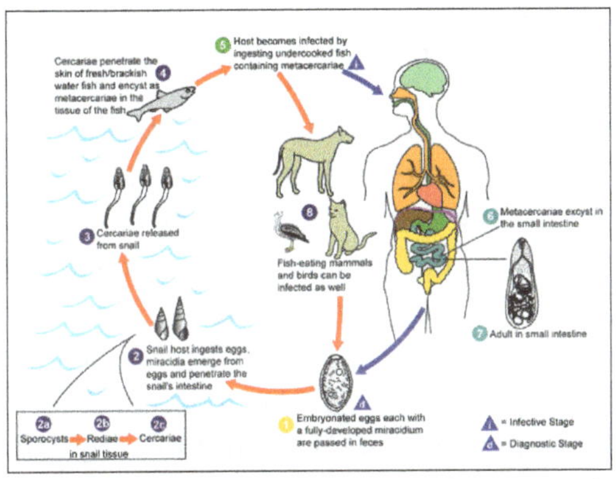

Life cycle of the digenean *Metagonimus*, an intestinal fluke.

Diseases and Parasites in Salmon

Sample of pink salmon infected with Henneguya salminicola, caught off the Queen Charlotte Islands, Western Canada in 2009.

According to Canadian biologist Dorothy Kieser, the myxozoan parasite *Henneguya salminicola* is commonly found in the flesh of salmonids. It has been recorded in the field samples of salmon returning to the Queen Charlotte Islands. The fish responds by walling off the parasitic infection into a number of cysts that contain milky fluid. This fluid is an accumulation of a large number of parasites.

Henneguya and other parasites in the myxosporean group have a complex life cycle, where the salmon is one of two hosts. The fish releases the spores after spawning. In the *Henneguya* case, the spores enter a second host, most likely an invertebrate, in the spawning stream. When juvenile salmon migrate to the Pacific Ocean, the second host releases a stage infective to salmon. The parasite is then carried in the salmon until the next spawning cycle. The myxosporean parasite that causes whirling disease in trout, has a similar life cycle. However, as opposed to whirling disease, the *Henneguya* infestation does not appear to cause significant incapacitation of the host salmon — even heavily infected fish tend to return to spawn successfully.

According to Dr. Kieser, a lot of work on *Henneguya salminicola* was done by scientists at the Pacific Biological Station in Nanaimo in the mid-1980s, in particular, an overview report which states that "the fish that have the longest fresh water residence time as juveniles have the most noticeable infections. Hence in order of prevalence coho are most infected followed by sockeye, chinook, chum and pink." As well, the report says that, at the time the studies were conducted, stocks from the middle and upper reaches of large river systems in British Columbia such as Fraser, Skeena, Nass and from mainland coastal streams in the southern half of B.C. "are more likely to have a low prevalence of infection." The report also states "It should be stressed that *Henneguya*, economically deleterious though it is, is harmless from the view of public health. It is strictly a fish parasite that cannot live in or affect warm blooded animals, including man".

According to Klaus Schallie, Molluscan Shellfish Program Specialist with the Canadian Food Inspection Agency, "*Henneguya salminicola* is found in southern B.C. also and in all species of salmon. I have previously examined smoked chum salmon sides that were riddled with cysts and some sockeye runs in Barkley Sound (southern B.C., west coast of Vancouver Island) are noted for their high incidence of infestation."

Sea lice, particularly *Lepeophtheirus salmonis* and various *Caligus* species, including *C. clemensi* and *C. rogercresseyi*, can cause deadly infestations of both farm-grown and wild salmon. Sea lice are ectoparasites which feed on mucus, blood, and skin, and migrate and latch onto the skin of wild salmon during free-swimming, planktonic nauplii and copepodid larval stages, which can persist for several days. Large numbers of highly populated, open-net salmon farms can create exceptionally large concentrations of sea lice; when exposed in river estuaries containing large numbers of open-net farms, many young wild salmon are infected, and do not survive as a result. Adult salmon may survive otherwise critical numbers of sea lice, but small, thin-skinned juvenile salmon migrating to sea are highly vulnerable. On the Pacific coast of Canada, the louse-induced mortality of pink salmon in some regions is commonly over 80%.

Some Background

In 1984, infectious salmon anemia (ISAv) was discovered in Norway in an Atlantic salmon hatchery. Eighty percent of the fish in the outbreak died. ISAv, a viral disease, is now a major threat to the viability of Atlantic salmon farming. It is now the first of the diseases classified on List One of the European Commission's fish health regime. Amongst other measures, this requires the total eradication of the entire fish stock should an outbreak of the disease be confirmed on any farm. ISAv seriously affects salmon farms in Chile, Norway, Scotland and Canada, causing major economic losses to infected farms. As the name implies, it causes severe anemia of infected fish. Unlike mammals, the red blood cells of fish have DNA, and can become infected with viruses. The fish develop pale gills, and may swim close to the water surface, gulping for air. However, the disease can also develop without the fish showing any external signs of illness, the fish maintain a normal appetite, and then they suddenly die. The disease can progress slowly throughout an infected farm and, in the worst cases, death rates may approach 100 percent. It is also a threat to the dwindling stocks of wild salmon. Management strategies include developing a vaccine and improving genetic resistance to the disease.

Sea lice on the back of an Atlantic salmon.

In 1972, Gyrodactylus, a monogenean parasite, spread from Norwegian hatcheries to wild salmon, and devastated some wild salmon populations.

In the wild, diseases and parasites are normally at low levels, and kept in check by natural predation on weakened individuals. In crowded net pens they can become epidemics. Diseases and parasites also transfer from farmed to wild salmon populations. A recent study in British Columbia links the spread of parasitic sea lice from river salmon farms to wild pink salmon in the same river. The European Commission (2002) concluded "The reduction of wild salmonid abundance is also linked to other factors but there is more and more scientific evidence establishing a direct link between the number of lice-infested wild fish and the presence of cages in the same estuary." It is reported that wild salmon on the west coast of Canada are being driven to extinction by sea lice from nearby salmon farms. These predictions have been disputed by other scientists and recent harvests have indicated that the predictions were in error. Antibiotics and pesticides are often used to control the diseases and parasites.

Wild Salmon

According to Canadian biologist Dorothy Kieser, protozoan parasite *Henneguya salminicola* is commonly found in the flesh of salmonids. It has been recorded in the field samples of salmon returning to the Queen Charlotte Islands. The fish responds by walling off the parasitic infection into a number of cysts that contain milky fluid. This fluid is an accumulation of a large number of parasites.

Henneguya and other parasites in the myxosporean group have a complex lifecycle where the salmon is one of two hosts. The fish releases the spores after spawning. In the *Henneguya* case, the spores enter a second host, most likely an invertebrate, in the spawning stream. When juvenile salmon out-migrate to the Pacific Ocean, the second host releases a stage infective to salmon. The parasite is then carried in the salmon until the next spawning cycle. The myxosporean parasite that causes whirling disease in trout, has a similar lifecycle. However, as opposed to whirling disease, the *Henneguya* infestation does not appear to cause disease in the host salmon — even heavily infected fish tend to return to spawn successfully.

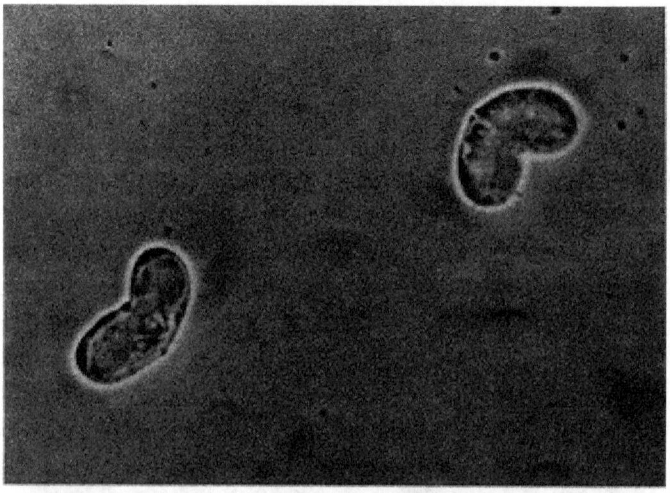

The myxosporean parasite *Ceratomyxa shasta* infects salmonid fish on the Pacific coast of North America.

According to Dr. Kieser, a lot of work on *Henneguya salminicola* was done by scientists at the Pacific Biological Station in Nanaimo in the mid-1980s, in particular, an overview report which states that "the fish that have the longest fresh water residence time as juveniles have the most noticeable infections. Hence in order of prevalence coho are most infected followed by sockeye, chinook, chum and pink." As well, the report says that, at the time the studies were conducted, stocks from the middle and upper reaches of large river systems in British Columbia such as Fraser, Skeena, Nass and from mainland coastal streams in the southern half of B.C. "are more likely to have a low prevalence of infection." The report also states "It should be stressed that *Henneguya*, economically deleterious though it is, is harmless from the view of public health. It is strictly a fish parasite that cannot live in or affect warm blooded animals, including man".

According to Klaus Schallie, Molluscan Shellfish Program Specialist with the Canadian Food Inspection Agency, "*Henneguya salminicola* is found in southern B.C. also and in all species of salmon. I have previously examined smoked chum salmon sides that were riddled with cysts and some sockeye runs in Barkley Sound (southern B.C., west coast of Vancouver Island) are noted for their high incidence of infestation."

Sea lice, particularly *Lepeophtheirus salmonis* and a variety of *Caligus* species, including *Caligus clemensi* and *Caligus rogercresseyi*, can cause deadly infestations of both farm-grown and wild salmon. Sea lice are ectoparasites which feed on mucous, blood, and skin, and migrate and latch onto the skin of wild salmon during free-swimming, planktonic *naupli* and *copepodid* larval stages, which can persist for several days. Large numbers of highly populated, open-net salmon farms can create exceptionally large concentrations of sea lice; when exposed in river estuaries containing large numbers of open-net farms, many young wild salmon are infected, and do not survive as a result. Adult salmon may survive otherwise critical numbers of sea lice, but small, thin-skinned juvenile salmon migrating to sea are highly vulnerable. On the Pacific coast of Canada, the louse-induced mortality of pink salmon in some regions is commonly over 80%.

Farmed Salmon

Atlantic salmon.

In 1972, Gyrodactylus, a monogenean parasite, spread from Norwegian hatcheries to wild salmon, and devastated some wild salmon populations. In 1984, infectious salmon anemia (ISAv) was discovered in Norway in an Atlantic salmon hatchery. Eighty percent of the fish in the outbreak died. ISAv, a viral disease, is now a major threat to the viability of Atlantic salmon farming. It is now the first of the diseases classified on List One of the European Commission's fish health regime. Amongst other measures, this requires the total eradication of the entire fish stock should an outbreak of the disease be confirmed on any farm. ISAv seriously affects salmon farms in Chile, Norway, Scotland and Canada, causing major economic losses to infected farms. As the name implies, it causes severe anemia of infected fish. Unlike mammals, the red blood cells of fish have DNA, and can become infected with viruses. The fish develop pale gills, and may swim close to the water surface, gulping for air. However, the disease can also develop without the fish showing any external signs of illness, the fish maintain a normal appetite, and then they suddenly die. The disease can progress slowly throughout an infected farm and, in the worst cases, death rates may approach 100 percent. It is also a threat to the dwindling stocks of wild salmon. Management strategies include developing a vaccine and improving genetic resistance to the disease.

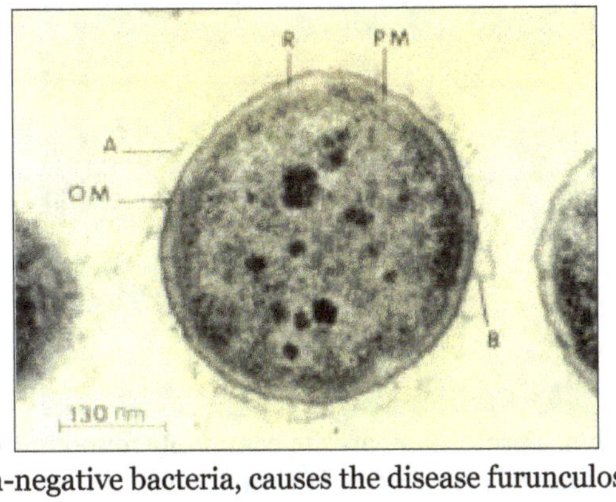

Aeromonas salmonicida, a Gram-negative bacteria, causes the disease furunculosis in marine and freshwater fish.

In the wild, diseases and parasites are normally at low levels, and kept in check by natural predation on weakened individuals. In crowded net pens they can become epidemics. Diseases and parasites also transfer from farmed to wild salmon populations. A recent study in British Columbia links the spread of parasitic sea lice from river salmon farms to wild pink salmon in the same river." The European Commission (2002) concluded "The reduction of wild salmonid abundance is also linked to other factors but there is more and more scientific evidence establishing a direct link between the number of lice-infested wild fish and the presence of cages in the same estuary." It is reported that wild salmon on the west coast of Canada are being driven to extinction by sea lice from nearby salmon farms. Antibiotics and pesticides are often used to control the diseases and parasites.

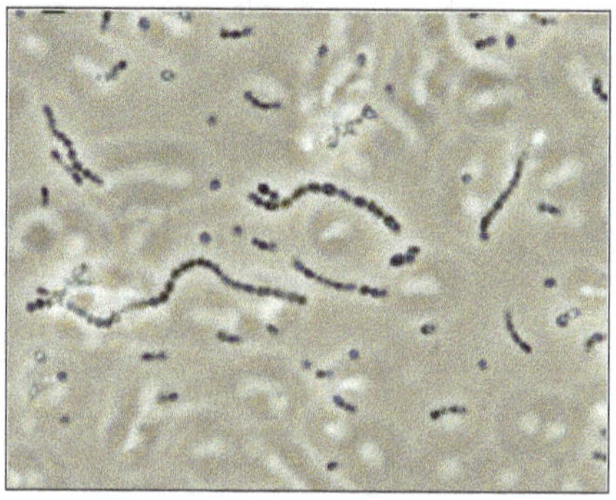

Streptococcus iniae, a Gram-positive, sphere-shaped bacteria caused losses in farmed marine and freshwater finfish of US$100 million in 1997.

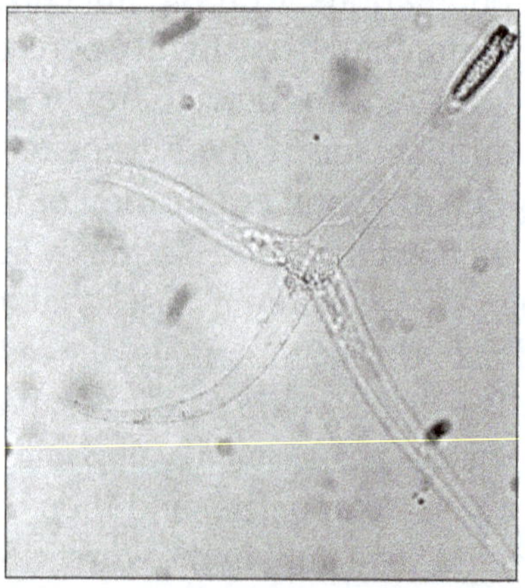

Myxobolus cerebralis, a myxosporean parasite, causes *whirling disease* in farmed salmon and trout and also in wild fish populations.

Henneguya salminicola, a protozoan parasite commonly found in the flesh of salmonids on the West Coast of Canada. Coho salmon.

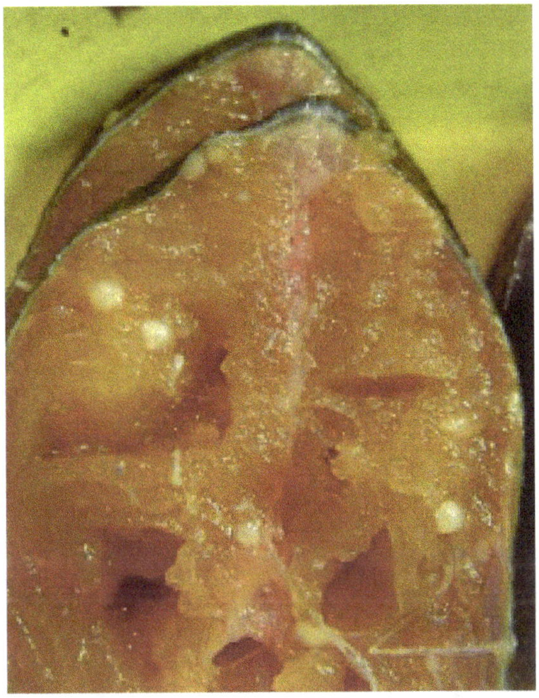

The parasite *Henneguya zschokkei* in salmon beard.

Columnaris in the gill of a chinook salmon.

- Sockeye salmon with gas bubble disease.

References

- Woo PTK (2011) Fish Diseases and Disorders: Volume 3: Viral, Bacterial and Fungal Infections Cabi Series. ISBN 9781845935542
- Fairfield, T (2000). A commonsense guide to fish health. Woodbury, N.Y: Barron's Educational Series. ISBN 0-7641-1338-0

- Exell A, Burgess PH, Bailey MT. A-Z of Tropical Fish Diseases and Health Problems. New York, N.Y: Howell Book House. ISBN 1-58245-049-8
- Fairfield, T (2000). A commonsense guide to fish health. Woodbury, N.Y: Barron's Educational Series. ISBN 0-7641-1338-0
- Moyle, PB and Cech, JJ (2004) Fishes, An Introduction to Ichthyology. 5th Ed, Benjamin Cummings. ISBN 978-0-13-100847-2
- Woo PTK (1995) Fish Diseases and Disorders: Volume 1: Protozoan and Metazoan Infections Cabi Series. ISBN 9780851988238

5

Aquaponics: Types and Approaches

Aquaponics is a form of aquaculture in which the nutrient rich aquaculture water is fed to plants grown in a soil-less environment. It can be considered as a combination of aquaculture and hydroponics. Algaeponics, vermiponics, aquaponic systems, aquaponic fish, aquaponic vegetables, bio filters, etc. are some of the aspects related to aquaponics which will be explained elaborately in this chapter.

Aquaponics is a sustainable method of raising both fish and vegetables. It is popular with individuals, entrepreneurs, educators, missions and governments. Furthermore, with this type of indoor farming, you grow substantially more food with less water, land and labor than traditional agriculture.

Aquaponics is a form of agriculture that combines raising fish in tanks (recirculating aquaculture) with soilless plant culture (hydroponics). In aquaponics, the nutrient-rich water from raising fish provides a natural fertilizer for the plants and the plants help to purify the water for the fish. Aquaponics can be used to sustainably raise fresh fish and vegetables for a family, to feed a village or to generate profit in a commercial farming venture, year 'round, in any climate.

Aquaponics is a great example of year 'round, indoor farming. It can be done anywhere, providing fresh local food that is free of pesticides, herbicides and chemical fertilizers. It is safe, easy and fresh.

- Home: A home food production system can grow hundreds of pounds of fish and all the fresh vegetables that a family needs.
- School: In education applications, aquaponics is an excellent model of nature's biological cycles.
- Food Banks and Missions: Feed people in need: Aquaponics is an ideal way to provide fresh, nutritious food to those in need.
- Commercial: Commercially, aquaponics is a rapidly growing industry as entrepreneurs realize that aquaponics and controlled environment agriculture can provide high quality, locally-grown fresh food on a 'year round basis. Large commercial aquaponic farms are providing fresh food to grocery store chains, hospitals and institutions. Micro Farm Aquaponics Systems are perfect for a family or retirement business.

Natural and Sustainable

Aquaponics is a completely natural process that mimics all lakes, ponds, rivers and waterways on Earth. The only input into an aquaponics system is fish food. The fish eat the food and excrete waste, which is converted by beneficial bacteria to nutrients that the plants can use. In consuming these nutrients, the plants help to purify the water. You cannot use herbicides, pesticides or other harsh chemicals in an aquaponics system, making the fish and plants healthful and safe to eat.

Benefits of Aquaponics

- Use 1/6th of the water to grow 8 times more food per acre compared to traditional agriculture.
- All natural fertilizer sources from fish waste.
- No reliance on mined and manufactured fertilizers.
- Efficient, sustainable and highly productive.
- Produce is free of pesticides and herbicides.
- Fish are free of growth hormones and antibiotics.
- Allows continuous production of food.
- Produces both a protein and vegetable crop.
- Integrated system is sustainable and earth-friendly.
- Eliminating soil eliminates soil borne diseases.

Aquaponics Operation

An aquaponic system isn't difficult to maintain but there are daily and periodic tasks that must be done to ensure a healthy system.

Fish Feeding

The availability of dry, species-specific fish food is quite new and is the result of the rapidly growing aquaculture industry. Today's specialized fish feeds provide precise amounts of protein, carbohydrates, fats, vitamins, amino acids and minerals. In a hobby or ornamental system it is fine to feed your fish only once a day. If food production is your goal, you'll want to maximize your efforts and feed more frequently. Most commercial growers will feed three times per day. Feeding by hand gives you the opportunity to observe the fish and their feeding habits. You can also use timed mechanical feeder or use an on-demand feeder so the fish can eat whenever they are hungry.

Plant Seeding, Rotation and Harvesting

When growing leafy crops that quickly mature, such as lettuce and herbs, you should plant frequently so you are assured of a continual harvest. With leafy crops, when you havest, you seed and transplant the same number of plants you harvest. This will provide continuous harvests.

Indeterminate varieties of many long term crops, tomatoes and melons, can produce for many months. Seeds for an aquaponic system are usually germinated in a small rockwool cube or in a container of loose perlite, vermiculite, a seedling mix or coco coir. Germination can be done within the aquaponic system or in a separate area designed to provide proper environmental conditions for germination. Once the seed has germinated and a seedling has developed, the small plant is transplanted into the plant bed.

Observation and Monitoring

Every day you should visually inspect the fish and the system. Make sure the water is flowing properly, the aeration system is working and the drains are free of debris. Watch to see that the fish eat vigorously and swim normally. Look for problems such as fungus, open soars, torn fins or discoloration and check the temperature and pH of the water. Observe the plants for pests and insects. Remove any dead plant matter.

Fish Harvesting

In a home food production system, you will most likely harvest fish as you want them, once they've matured. Commercial operations will harvest based on market demand and production. Farm raised fish always taste the best if they've been purged for several days prior to harvesting. Ideally, the fish should be removed from the main system and held without feed for several days prior to harvest.

Water Quality Testing

Anyone serious about aquaponics should invest in a water test kit, which will enable you to measure and keep track of pH, ammonia, nitrite, nitrate, dissolved oxygen, alkalinity and water hardness. Most kits sold are easy to use and are based on color changes in the sample being tested. Meters that measure dissolved oxygen, temperature and other factors are also available.

Cleaning Filters and System

The key to a healthy system is keeping it clean, removing any dead or unhealthy plants or fish, maintaining filters and emptying the clarifier if you are using one.

Aquaponics Biofilter

A biofilter is a place for the bacteria to colonize. It provides large surface areas, pH, dissolved oxygen levels, and proper temperature.

Biofilters are quite simple to set up, and they consist of a tank connected to the air pump and some substrates that provide as much as possible a big surface area for the nitrifying bacteria to grow. The more the surface area for the bacteria to develop, the better it is because it means a more efficient nitrification process.

Biofiltration

Biofiltration is the conversion of ammonia and nitrite into nitrates by the beneficial bacteria. Ammonia and nitrites are toxic to aquaponics, even at low concentrations. However, plants need the nitrates to grow, so a biofilter needs to be installed to house most living bacteria. The water movement within a biofilter will break down the very fine solids not captured by the clarifier which prevents a waste build up on the plant roots in the NFT aquaponics system. Good biofiltration is also essential for the chemical stability of your aquaponics system. Separate bio filters are unnecessary for the media-based aquaponics system because the grow beds themselves are perfect biofilters.

Need of a Biofilter

In aquaponics, if the surface area in your grow media is not enough for the bacteria to colonize, you need to add more surface area. This is why you need to have a biofilter. The biofilter will form an important part of your system by ensuring that plants have enough nutrients to grow healthy while cleaning the water for the fish to live.

Your whole aquaponics system is a biofilter. However, if you are using the DWC, Vertical, and NFT aquaponics system to grow your plants, you will need a biofilter. In these types of aquaponics systems, you can't rely on the natural process of bacterial conversion because of a lack of adequate surface area. A media-based or raft aquaponics system does not usually need a separate biofilter because the raft, growing media (expanded clays, gravel, or lava rock), tank walls, and other surface areas provide enough area for the bacteria to colonize.

Benefits of using Biofilter in Aquaponics

- It provides more surface area for the beneficial bacteria to live and helps convert ammonia and nitrites into nitrates.
- Biofilter assists in aeration and nitrification processes in the NFT, DWC, and Vertical aquaponics systems.
- Aid in the chemical stability of your aquaponics system.
- Very inexpensive and easy to build.
- Very easy to use and require very little maintenance.

Important Factors to Consider Before using a Biofilter

Before you decide to use biofilter in your aquaponics system, you must be aware of these considerations:

- Remove the solids first: Biofilters are not for large solid fish wastes. Solids must be screened because this will possibly block the pipes. Once these solids accumulate in your biofilter, this will create areas known as aerobic zones that will attract bacteria and reduce your bio filter's surface area. This will reduce your bio filter's efficiency, which will result in a high ammonia level, which is not suitable for fish and plants. That is why it is important that you add a standard filter that will remove the solid.

- Bacteria needs air: In an aquaponics system, the bacteria are exposed to the air. Adding a biofilter in your aquaponics system will contain the bacteria in the filter. So you have to make sure that your filter has enough air supply to ensure that the bacteria are not deprived of air.
- Fish density: If you are running a low fish density, then you don't need a biofilter. Low quantities of fish will not produce enough waste to justify the use of a biofilter. The plant's roots will provide enough surface area for the bacteria to convert ammonia into nitrates.

Working Principles of Aquaponics Biofilters

The biofilter is the extension of the surface area of your aquaponics system. It is an essential part of your aquaponics system because it ensures that your plants can access the nitrates that they need to grow while cleaning the water for the fish. The bacteria will attach to the surface of your biofilter media. Once they are established, they do their job and help convert ammonia and nitrites into nitrates. Biofiltration in aquaponics happens in three main steps:

- The air pump pumps the water out of the fish tank and into the biofilter.
- Inside the biofilter, the water goes through the nitrification process. The beneficial bacteria help convert ammonia and nitrites into nitrates.
- The water full of nutrients flows out of the biofilter and into the plants, where the plant roots absorb the nutrients while cleaning the water before it goes back into the fish tank.

This process is for typical aquaponics set up. The process or steps may vary depending on your biofilter design.

Different Kinds of Biofilter

There are several aquaponics bio filters that you can use in your aquaponics system; these are:

- Moving bed filter: This biofilter is used in a floating raft system where there is not much surface area for the bacteria.
- Static filter: These are tray-like filters that can be slid into a separate container next to the fish tank before the plants. These provide a large area for the bacteria to collect on. This filter needs to be positioned after the solid filter before it reaches the plants, allowing the water to be filtered before it reaches the plants.
- Drip filter: This filter allows water to drop in from the top. The water passes through the filter box, full of gravel, oyster shells, or another similar bio medium with lots of surface area. This biofilter also needs to be positioned where the water is filtered before it gets to the plants. As the water moves through the filter, the ammonia is converted to nitrates before it is pumped back to the plants.

Hydroponics

Hydroponics is the growing of plants in a liquid nutrient solution with or without the use of artificial media. Commonly used mediums include expanded clay, coir, perlite, vermiculite, brick shards, polystyrene packing peanuts and wood fiber. Hydroponics has been recognized as a viable method of producing vegetables (tomatoes, lettuce, cucumbers and peppers) as well as ornamental crops such as herbs, roses, freesia and foliage plants. Due to the ban on methyl bromide in soil culture, the demand for hydroponically grown produce has rapidly increased in the last few years.

Advantages

- It can be used in places where in-ground agriculture or gardening is not possible (for example, dry desert areas or cold climate regions).
- More complete control of nutrient content, pH and growing environment.
- Lower water and nutrient costs associated with water and nutrient recycling.
- Faster growth due to more available oxygen in root area.
- Elimination or reduction of soil related insects, fungi and bacteria.
- Much higher crop yields.
- No weeding or cultivation required.
- Some crops, such as lettuce and strawberries, can be lifted from ground level to a much better height for planting, cultivation and harvesting. This gives much better working conditions and hence lowers labor costs.

- Crop rotation/fallowing is not necessary.
- Transplant shock is reduced.

Disadvantages

- Initial and operational costs are higher than soil culture.
- Skill and knowledge are needed to operate properly.
- Some diseases like Fusarium and Verticillium can spread quickly through the system. However, many varieties resistant to the above diseases have been bred.

Growing Systems

Hydroponic systems can either be liquid or aggregate. Liquid systems have no supporting medium for the plant roots; whereas, aggregate systems have a solid medium of support. Hydroponic systems are further categorized as open (once the nutrient solution is delivered to the plant roots, it is not reused) or closed (surplus solution is recovered, replenished, and recycled).

Liquid Hydroponic System

They are closed systems:

- Nutrient Film Technique (NFT): Plants are placed in a polyethylene tube that has slits cut in the plastic for the roots to be inserted. Nutrient solution is pumped through this tube.
- Floating Hydroponics: Plants are grown on a floating raft of expanded plastic.
- Aeroponics: Plant roots remain suspended in an enclosed growing chamber, where they are sprayed with a mist of nutrient solution at short intervals, usually every few minutes.

Aggregate Hydroponic System

- Open system:
    - Rockwool Culture: It is the most widely used medium in hydroponics. Rockwool is ground-up basalt rock that is heated then spun into threads making wool. It is very light and is often sold in cubes. Rockwool can hold water and retain sufficient air space (at least 18 percent) to promote optimum root growth.
    - Sand Culture.
- Closed system:
    - Gravel.
    - NFT and Rockwool: Plants are established on small rock wool slabs positioned in channels containing recycled nutrient solution.

These systems are further categorized into:

- Passive systems.
- Active systems.

Passive systems use a wick and growing media with very high capillary action. This allows water to be drawn to the plant roots. The Wick System is by far the simplest type of hydroponic system.

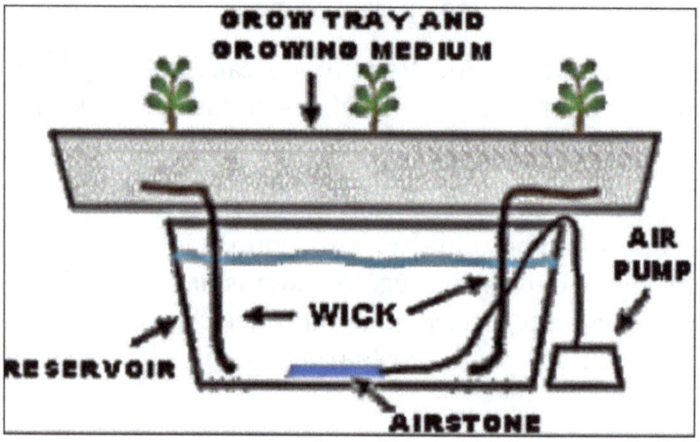

Figure: The wick system.

Active systems work by actively passing a nutrient solution over your plants roots. Examples include:

- The water culture system is the simplest of all active hydroponic systems. The platform that holds the plants is usually made of Styrofoam and floats directly on the nutrient solution. An air pump supplies air to the air stone that bubbles the nutrient solution and supplies oxygen to the roots of the plants.

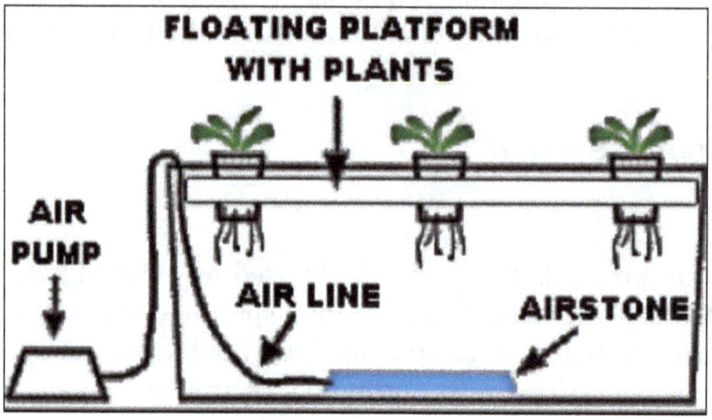

Figure: The water culture system.

- The Ebb and flow system works by temporarily flooding the grow tray with nutrient solution and then draining the solution back into the reservoir. This action is normally done with a submerged pump that is connected to a timer. The timer is set to come on several times a day, depending on the size and type of plants, temperature, humidity and the type of growing medium used.

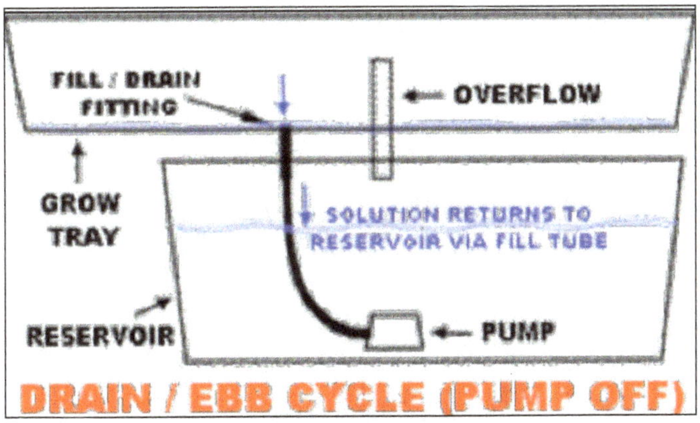

Figure: The Ebb and Flow system.

- Drip Systems are probably the most widely used type of hydroponic system in the world. A timer controls a submersed pump. The timer turns the pump on and nutrient solution is dripped onto the base of each plant by a small drip line.

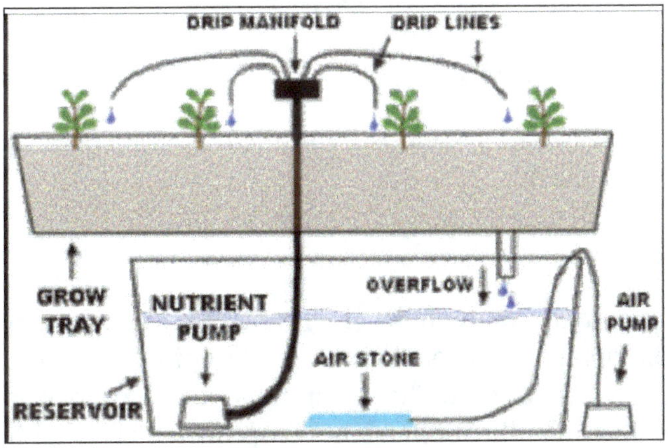

Figure: The Drip system.

- NFT Systems have a constant flow of nutrient solution, so no timer is required for the submersible pump.

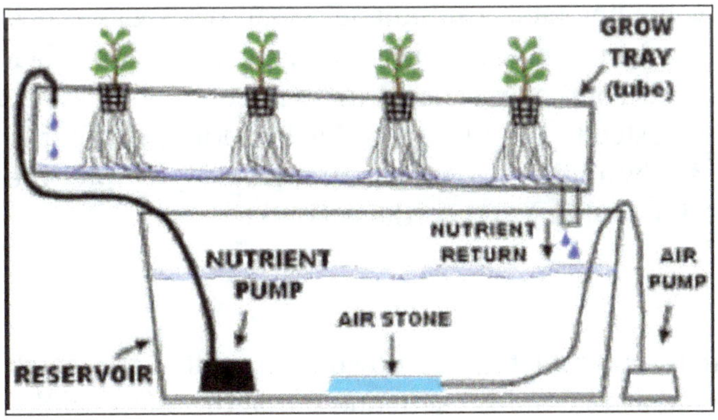

Figure: The NFT system.

- The Aeroponic System is probably the most high-tech type of hydroponic gardening. A timer controls the nutrient pump much like other types of hydroponic systems, except the aeroponic system needs a short cycle timer that runs the pump for a few seconds every couple of minutes.

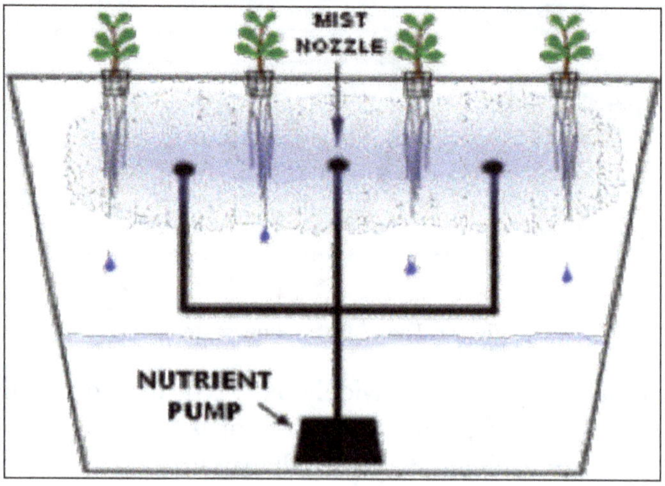

Figure: The Aeroponic system.

Nutrient Management Techniques

The major disadvantage of a closed system is the difficulty of nutrient management. Four main techniques are commonly utilized:

Technique 1

Water addition, pH and electrical conductivity (EC) control are all automatic. The pH is a measure of the acidity of the substrate and controls the availability of mineral nutrients; whereas, the EC gives an estimate of the nutrient content. The recommended pH for hydroponic culture is between 5.0 and 6.0 because overall availability of nutrients is optimized at a slightly acidic pH, and the EC level should be 1.5 to 3 dS m^{-1}.

Technique 2

The water makeup of the holding tank is automatic, usually by float valve, i.e. the tank level is held steady. Here both water and nutrients are being taken, but only water is being replaced. Therefore, the EC will fall until the tank solution is brought up to strength by nutrient addition. The EC is periodically checked and adjusted to the required value by adding nutrient to the tank by hand. The pH is adjusted if necessary by adding acid (dilute sulfuric acid) to lower the pH or an alkali (dilute sodium hydroxide (NaOH) solution) to raise the pH.

Technique 3

The holding tank is partly or completely run down then refilled as a batch by adding water and/or nutrient. The important aspect of this technique is that the effects of the addition are checked.

Technique 4

The holding tank is partly or completely run down then refilled using a standard strength nutrient solution. However, the resultant EC in the system is not checked or adjusted. This technique can lead to disaster. Nutrient Solutions for Hydroponics Commercial hydroponic growers need a more accurate control of the components in a nutrient solution to achieve commercial success. Numerous 'recipes' for hydroponic solutions are available. Many use different combinations of chemicals to reach similar total final compositions.

Table: Major element and micronutrient ionic forms and normal concentration range found in most nutrient solutions.

Element	Ionic Form	Concentration Range mg/L, ppm
Major elements		
Nitrogen (N)	NO^-_3, NH^+	100 to 200
Phosphorus (P)	HPO^{2-}_4, $H_2PO^-_4$	30 to 15
Potassium (K)	K^+	100 to 200
Calcium (Ca)	Ca^{2+}	300 to 300
Magnesium (Mg)	Mg^{2-}_4	30 to 80
Sulfur (S)	SO^{2-}_4	70 to 150
Micronutrients		
Boron (B)	BO^{3-}_3	0.03
Chlorine (Cl)	Cl^-	-
Copper (Cu)	Cu^{2+}	0.01 to 0.10
Iron (Fe)	Fe^{2+}, Fe^{3+}	20 to 12
Manganese (Mn)	Mn^{2+}	0.5 to 2.0
Molybdenum (Mo)	Mo	0.05
Zinc (Zn)	Zn^{2+}	0.05 to 0.50

Soil Versus Hydroponics

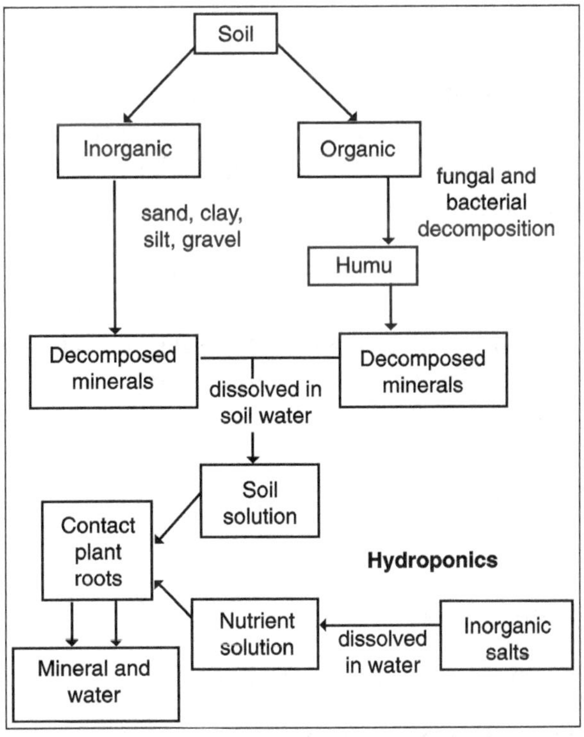

Figure: Origin of essential elements in soil and hydroponics.

There is no physiological difference between plants grown hydroponically and those grown in soil. In soil, both the organic and inorganic components must be decomposed into inorganic elements before they are available to the plant. These elements adhere to the soil particles and are exchanged into the soil solution where they are absorbed by plants. In hydroponics, the plant roots are moistened with a nutrient solution containing the elements. The subsequent processes of mineral uptake by the plant are the same.

Nutrition and Hydroponics

There is no conclusive evidence that produce grown hydroponically are more nutritious or healthier than produce grown by any other method, although some small studies indicate that it may be possible. Many countries, such as Holland, do not differentiate whether the produce has been grown by hydroponics or by any other methods; they just concentrate on the quality of the produce. Produce quality is more likely to remain consistent in hydroponic systems, as plants are often less stressed than in other systems.

Hydroponics in the Home Garden

For the hobbyist, hydroponics is a highly technical specialized form of agriculture that has an ever expanding application to everyday use. At the same time, home hydroponic gardens have become so simple that people without a green thumb can be successful in growing flowers and vegetables in their homes. Many of the advantages of commercial hydroponics also apply to the home and hobby gardeners. In many industries, small systems are developed, and then expanded for large scale use. Conversely, in hydroponics, large scale commercial production is becoming more common, while it is more of a challenge to make smaller systems economically feasible.

Planning Commercial Hydroponics

Hydroponic systems are only one of the options available when you are considering whether to grow a crop. Planning for a commercial business should, therefore, follow the normal sequence for considering any horticultural enterprise. Don't overlook soil growing. You need a worthwhile reason to use a hydroponic system instead of soil.

If you do choose to go with hydroponics, you should evaluate the advantages and disadvantages of each type of production system for your crop of interest. For short term crops such as lettuce, the common choice is recirculating NFT or flood and drain gravel channels. For longer term crops or those very vulnerable to root disease, the common choice is non-recirculating, media-based systems.

Advantages and Disadvantages

Advantages

- You can grow anywhere.
- Uses 20 times less water than soil based gardening.
- Your environment is sterile, which means no pesticides.

- You'll use 20% less space for growing.
- The system water can be reused, allowing you to conserve water.
- You can have complete control over nutrient balance by using Dyna-Gro Nutrition Solutions.
- No soil setup and testing hassles.
- Harvesting is easier.
- No mulching, tilling, changing of soil and weeding.
- You can grow year round if indoors.

Disadvantages

- Putting together a hydroponic system isn't cheap.
- Constant monitoring is required.
- Hydroponic systems are vulnerable to power outages. In the event of a power outage that outlasts your generators you will be manually watering your garden.
- Micro-organisms that are water-based can creep in rather easily.
- Growing a hydroponic garden demands technical expertise.
- Production is limited compared to field conditions.
- If a disease appears, all plants in the system will be affected.
- Without soil to serve as a buffer if the system fails plant death will occur rapidly.

Algaeponics

Microalgae are unicellular photoautotrophs (ranging from 0.2 μm up to 100 μm) and are classified in various taxonomic groups. Microalgae can be found in most environments but are mostly found in aquatic environments. Phytoplanktons are responsible for over 45% of world's primary production as well as generating over 50% of atmospheric O_2. In general, there is no major difference in photosynthesis of microalgae and higher plants. However, due to their smaller size and the reduction in a number of internally competitive physiological organelles, microalgae can grow much faster than higher plants. Microalgae can also grow under limited nutrient conditions and have the ability to adapt to a wider range of environmental conditions. Most importantly, microalgal culture does not compete with food crop production regarding arable land and freshwater. Furthermore, microalgae can efficiently utilize inorganic nutrients from waste effluents. In general, microalgal biomass contains up to 50% carbon making them a perfect candidate for bio remediating atmospheric CO_2.

The increase in extensive worldwide agriculture and animal farming has resulted in significant increases in biologically available nitrogen and phosphorus entering the terrestrial biosphere. Crop and animal farming and sewage systems contribute significant amounts to these nutrient loads. The infiltration of these nutrients into water streams can cause massive environmental issues such as harmful algal blooms and mass fish mortality. For instance, in the USA, nutrient pollution from agriculture is acknowledged as one of the major sources of eutrophication. Controlling the flow of nutrients from farming operations into the surrounding environment results in both, technical and economic challenges that must be overcome to reduce such effects. There have been various successful processes developed to treat waste effluent with high organic loads. However, almost all of these methods are not very effective in removing inorganic elements from water. Furthermore, some of these methods are rather expensive to operate. One simple method for treating organic waste is anaerobic digestion (AD). The AD process is well understood and when operated efficiently, it can convert over 90% of the wastewater organic matters to bio-methane and CO_2. The methane can be used to generate electricity and the generated heat can be used for various additional purposes. However, the AD process results in creating an anaerobic digestion effluent (ADE) which is very rich in inorganic phosphate and nitrogen as well as high COD (carbon oxygen demand). In certain locations, this effluent can be treated using microalgae and macroalgae.

Algal Growth Systems

Since the United Nations committee recommended that conventional agricultural crops be supplemented with high-protein foods of unconventional origin, microalgae have become natural candidates. The first microalgal cultivation was achieved though in 1890 by culturing Chlorella vulgaris. Due to the fact that microalgae normally divide at a certain time of the day, the term cyclostat was developed in order to introduce a light/dark (circadian) cycle to the culture. The large-scale culturing of microalgae and the partial use of its biomass especially as a base for certain products such as lipids was probably started seriously as early as 1953 with the aim of producing food from a large-scale culture of Chlorella. Typically, algae can be cultured in liquid using open ponds, closed photo bioreactors, or a combination of these systems. Alga can also be cultured as biofilms.

Closed Photo Bioreactors

Closed algal cultures (photo bioreactors) are not exposed to the atmosphere but are covered with a transparent material or contained within transparent tubing. Photo bioreactors have the distinct advantage of preventing evaporation. Closed and semi-closed photo bioreactors are mainly used for producing high-value algal products. Due to the overall cost of operating expenditure (OPEX) and capital expenditure (CAPEX), closed photo bioreactors are less economical than open systems. On the other hand, there is less contamination and less CO_2 losses, and by creating reproducible cultivation conditions and flexibility in technical design, this makes them a good substitute for open ponds. Some of the closed systems' weaknesses can be overcome by:

- Reducing the light path.
- Solving shear (turbulence) complexity, reducing oxygen concentration.
- A temperature control system.

Closed photo bioreactors are mainly divided into carboys, tubular, airlift and plate photo bioreactors.

Open Ponds

Open ponds are most commonly used for large-scale outdoor microalgal cultivation. Major algal commercial production is based in open channels (raceways) which are less expensive, easier to build and operate when compared to closed photo bioreactors. In addition, the growth of microalgae meets fewer difficulties in open than closed cultivation systems. However just a few species of microalgae (e.g. Dunaliella salina, Spirulina sp., Chlorella sp.) have been grown successfully in open ponds. Commercial microalgal production costs are high, approximated to be between 4 and 20 $US/g^{-1}. Large-scale outdoor open pond commercial microalgal culture has developed over the last 70 years, and both still (unstirred) and agitated ponds have been developed and have been used on a commercial basis. The very large unstirred open ponds are simply constructed from natural water ponds with open beds that are usually less than 0.5 m in depth. In some smaller ponds the surface may be lined with plastic lining sheets. Unstirred open ponds represent the most economical and least technical of all commercial culture methods and have been commercially used for Dunaliella salina β-carotene production in Australia. Such ponds are mainly limited to growing microalgae which are capable of surviving in poor conditions or have a competitive advantage that allows them to outgrow contaminants such as protozoa, unwanted microalgae, viruses, and bacteria. Agitated ponds on the other hand have the advantage of a mixing regime. Most agitated ponds are either (a) circular ponds with rotating agitators or (b) single or joined raceway ponds.

Circular cultivation ponds have primarily been used for the large-scale cultivation of microalgae especially in South East Asia. The circular ponds up to 45 m in diameter and usually 0.3–0.7 m in depth are uncovered, but there are some examples which are covered by glass domes. The low shear stresses that are required for microalgae production are produced in these systems particularly in the centre of the pond, and this is a distinct advantage of these kinds of systems. Some disadvantages include expensive concrete structures, inefficient land use with large footprints, difficulties in controlling the movement of the agitating device and the added cost in supplying CO_2.

Paddlewheel-driven raceways are the most common commercial microalgal cultivation system. Raceways are usually constructed in either a single channel or as linked channels. Raceways are usually shallow (0.15 to 0.25 m deep), are constructed in a loop and normally cover an area of approximately 0.5 to 1.5 ha. Raceways are mostly used and recommended for the major commercial culturing of three species of microalgae including Chlorella, Spirulina and Dunaliella. A high risk of contamination and low productivity, resulting mainly from poor mixing regimes and light penetration, are the main disadvantages of these open systems. In raceways, biomass concentrations of up to 1000 mg dry weight.L^{-1} and productivities of 20 g dry weight.m^{-2}.d^{-1} have been shown to be possible.

The price of microalgal production makes economic achievement highly dependent on the marketing of expensive and exclusive products, for which demand is naturally restricted. Raceways are also the most used cultivation system used for treating wastewater.

Solid Cultivation

An alternative microalgal cultivation method is immobilizing the cells in a polymer matrix or attaching them to the surface of a solid support (biofilm). In general, the biomass yield of such biomass cultures are at least 99% more concentrated than liquid-based cultures. Dewatering is one of the most expensive and energy-intensive parts of any mass algal production. The main advantage of biofilm growth is the potential of reducing the dewatering process and the related energy consumption and thus costs. Biofilm cultivation can also increase cellular light capture, reduce environmental stress (e.g. pH, salinity, metal toxicity, very high irradiance), reduce the cost of production and reduce nutrient consumption. Solid-based cultivation methods can be used for treating wastewater (nutrient and metal removal). There are three main methods for biofilm cultivation: 100% directly submerged in medium, partially submerged in medium and using a porous substrate to deliver the nutrients and moisture from the medium to the cells.

Algal Growth Nutrient Requirements

A number of inhibitory physical, chemical and biological factors can inhibit high microalgal production. A basic knowledge of the critical growth limitations is probably the most essential factor before applying any microalgae to any process. Light is by far the most important limiting factor affecting the growth of any alga? Temperature is also a critical factor for mass algal production. However, these variables are difficult to control. Next to light and temperature, nutrients are the most important limiting factor affecting the growth of any alga and each microalgal species tends to have its own optimum nutrient requirements. The most important nutrients are nitrogen, phosphorus and carbon. Most algae respond to N-limitation by increasing their lipid content. For example, Shifrin and Chisholm reported that in 20 to 30 species of microalgae that they examined, the algae increased their lipid content under N-deprivation. Phosphorus is also an important nutrient required for microalgal growth as it plays an essential role in cell metabolism and regulation, being involved in the production of enzymes, phospholipids and energy-supplying compounds. Brown and Button's studies on green alga Selenastrum capricornutum showed an apparent growth limitation when the phosphate concentration of the medium was lower than 10 nM. CO_2 is also a critical nutrient for achieving high algal productivity. For example, if additional CO_2 is not added to the algal culture, the average productivity can be reduced by up to 80%. However, the addition of CO_2 to algal ponds is rather costly. The most economical way for introducing CO_2 to a culture media is the direct transfer of the gas into the media by bubbling through sintered porous stones or using pipes under submerged plastic sheets as CO_2 injectors. Unfortunately, in all of these methods there is still high loss of CO_2 to the atmosphere because of the short retention time of the gas bubbles in the algal suspension.

Although adding N, P and C is critical, other nutrients also affect microalgal growth and metabolism. A lack of other nutrients, such as manganese (Mn) and various other cations (Mg^{2+}, K^+ and Ca^{2+}), is also known to reduce algal growth. Trace elements are also critical for microalgal growth and some microalga also require vitamins for their growth. One effective and inexpensive way of supplying nutrients is by combining algal culture and wastewater treatment.

Algae and Wastewater Treatment

With an increase in environmental deterioration and a greater necessity to generate alternative food and energy sources, there is the impetus to explore the feasibility of biological wastewater treatments coupled with resource recovery. Microalgal wastewater treatments have been particularly attractive, due to algal photosynthetic activities, where light is transferred into profitable biomass. Under certain conditions, wastewater-grown microalgal biomass can be equivalent or superior in biomass production to higher plant species. Thus, the process can transform a waste product into useful products (e.g. animal feed, aquaculture feed, bio-fertilizer and bioenergy). Thus, the waste effluent is no longer a negative waste product, but it becomes a valuable substrate for producing important substances and successful microalgal wastewater bioremediation has been reported for over half a century. Algal phytoremediation indeed provides an environmentally favourable solution for the treatment of wastewater as it can utilize organic and inorganic nutrients efficiently. Microalgal cultures hold an enormous potential for the later steps of wastewater treatment, especially for reducing 'N', 'P' and 'COD'. Moreover, the added ability of microalgae to grow via different nutritional conditions such as photoautotrophic, mixotrophic and heterotrophic conditions also enhances its capabilities in removing various different types of pollutants and chemicals from aqueous matrices. The ability of microalgae in sequestrating carbon (CO_2) allows CO_2 bioremediation. The synchronized algal-bacteria relationship established is also ideally synergetic for the bioremediation of wastewater. Through photosynthesis, microalgae provide oxygen required by aerobic bacteria for the mineralization of organic matter as well as the oxidation of NH_4^+. In return, the bacteria supply carbon dioxide for the growth of microalgae, significantly reducing the amount of oxygen required for the overall wastewater treatment process. In general, waste effluents with low carbon to nitrogen ratios are fundamentally suited to the growth of photosynthetic organisms. Most importantly, the microalgal domestic and agricultural wastewater treatments is an attractive option since the technology is relatively easy and they require very low energy compared to the standard of effluent treatment. Optimization of microalgal wastewater treatment in large-scale raceway ponds is appealing since it combines the effective treatment of a harmful waste product and the production of potentially valuable protein-rich algal biomass. Figure below summarizes a closed loop system for treating any organic waste by combination of anaerobic digestion and algal cultivation.

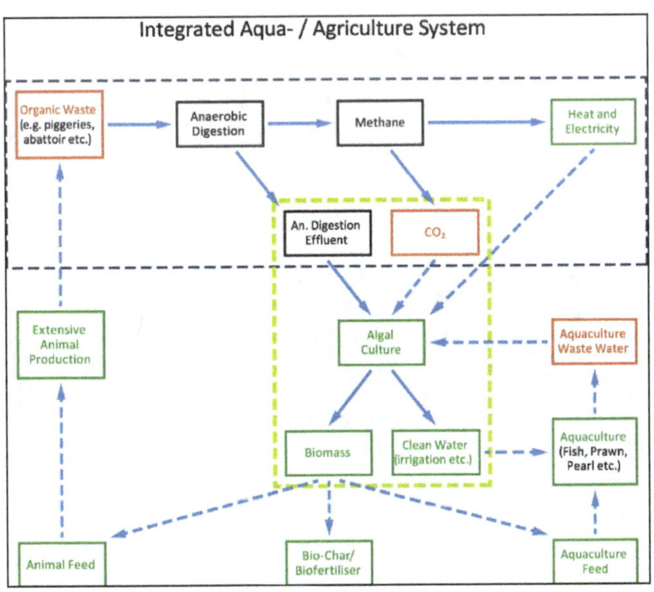

Figure: Integrated process system to use algal culture for treating organic waste and potential end users.

Algae and Aquaponics

Microalgae in aquaculture and in aquaponic systems is most often seen to be a nuisance as they can restrict water flows by clogging up pipes, consume oxygen, may attract insects, reduce water quality and when decomposing can deplete oxygen. However, an experiment by Addy et al. shows that algae can improve water quality in an aquaponic system, help control pH drops related to the nitrification process, generate dissolved oxygen in the system, 'produce polyunsaturated fatty acids as a value-added fish feed and add diversity and improve resilience to the system'. One of the 'holy grails' of aquaponics is to produce at least part of the food that is fed to the fish as part of the system and it is here that research is required in producing algae that could be grown with part of the aquaponics water, most probably in a separate loop, which can then be fed as part of the diet to the fish.

Aeroponics

Agricultural land is reducing day by day due to growing population needs. It is a challenge for farmers to provide food to such a great population. By natural growing process, the crops cultivated are not sufficient. In order to increase the production rate, farmers use chemicals. No doubt the chemicals increase the productivity but on the other hand degrade the nutritious value of food. These chemicals lead to many skin problems and diseases to living being and disturb the entire food chain. To overcome this major problem, an aeroponic system may be used.

The term Aeroponics means to cultivate plants without using soil and water as medium; by maintaining all the parameters essential for growth of plants. The parameters are temperature, humidity, pH & electrical conductivity of nutrient solution etc. resulting in a conditioned environment. The benefit of using Aeroponics is plant's healthier growth and nutritious fruit by consuming fewer amounts of nutrients and water. By adopting this technique fresh and healthier fruits can be produced throughout the year. In the recent decade, Aeroponic system is used for growing potatoes and for potato seed production.

Under these circumstances, controlled environment has a strong potential to improve plants' developmental stages, health, and growth. In the recent decade, aeroponic system is applied intensively for the purpose of growing potato in order to produce disease-free potato seed and in order to have a pesticides-free cultivation environment.

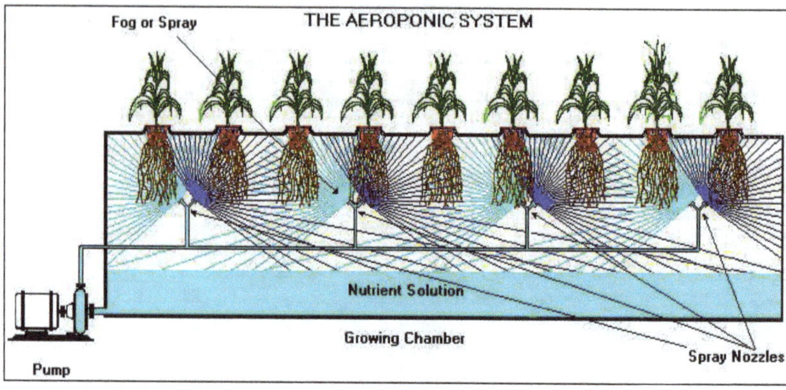

Figure: Basic Diagram of Aeroponic system.

The aeroponic system has more advantages than hydroponic in terms of spraying high air content in the nutrient solution to provide oxygen to plant roots. The focus of attention variables in the aeroponic system research are:

- Micro-environment (temperature, humidity, pH).
- The effectiveness of nutrition (spraying /fogging).

Aeroponic system is an endless process in a confined space and therefore it cuts down agricultural labour. A monitoring and control system intended for water and nutrients distribution has been designed to support the optimal application of aeroponic cultivation system for plant production.

Benefits of Aeroponic Technology

Aeroponic technology has several advantages as shown below:

- Aeroponic technology gives high yield with less space requirements.
- Plants can be grown close together.
- Fruits produced from the system are easier to harvest.
- Using this technique, fruits can be grown at zero gravity i.e at moon stations.
- Local food production can cuts down transportation costs.
- This technology saves water as it cuts down water consumption by 98 percent.
- Fresh and healthy plants can be grown at homes; indoors or at roof top.
- Nurseries can propagate seeds and cuttings into healthy, harvestable plants in a fraction of time of traditional methods.
- Plants and root growth study in laboratories is easier for students and researchers.
- Plants consume more oxygen under aeroponic conditions; more oxygen equals more plant growth.
- Planting and harvesting can be done throughout the year.
- Due to clean and sterile growing conditions, plant diseases and infections reduce up to a great extent.
- In aeroponic system, plant roots have proper space to grow well. So they don't stretch or wilt. Plants can be shifted to any growing media system without any transplantation shock after root development.
- Aeroponics systems can reduce water usage by 98 percent, fertilizer usage by 60 percent, and pesticide usage by 100 percent, all while maximizing crop yields.
- Aeroponic plants are potentially healthier and nutritious.
- Power loss for a small period does not cause any damage to plants.

Drawbacks of Aeroponic Technology

Like every coin has two sides, similarly every technology has major benefits as well as some negligible effect limitations. These limitations of aeroponic system are as follows:

- Power loss for a long time period may cause irreversible damage.
- Initially some training is required for system maintenance.
- Sanitary conditions are required to be maintained regularly.
- Initial cost of system is high.

Vermiponics

Vermiponics is a novel way to grow plants that uses the nutrients from worm castings (specifically from worm-tea) to grow plants in a soil-less (or hydroponic) environment. Vermiponics is similar to aquaponics (which is also soil-less) but instead of the nutrients being provided by fish waste that is broken down by bacteria, the nutrient solution in vermiponics is made from worm-tea instead.

Benefits of Vermiponics

Less Water Usage than Aquaponics

When you fertilize with betta fish or other fish, you need enough water for the fish to live in, as well as the water than runs nutrients through your indoor garden. And you will need to refresh the water in the fish tank every week. Depending on the size of your hydroponic setup, you could be going through quite a bit of water. But the worms will live in soil and compost with water added to keep everything moist. The vermiponic setup uses much less water than traditional aquaponics.

More Nutrients for Plants

Worm castings are a known natural fertilizer. They provide nitrogen and trace minerals. In fact, even if you don't get into vermiponics for your hydroponic garden, you can purchase earthworm castings as an organic fertilizer. Another option is to make worm tea from the castings. In this way you are utilizing the benefits of natural nutrients from worms in your indoor garden but keeping the vermicomposting (worm composting) separate from your hydroponic setup.

Great for Smaller Setups

Worms are quite small. Although you could make a small betta fish tank with one fish, worms are a terrific alternative for smaller indoor gardens. And aquaponics growers have found that worms work for their larger garden setups as well.

Easier than Caring for Fish

Fish require space and specialized food, as well as access to oxygen. Although worms also require oxygen, they utilize it more efficiently than fish, meaning they can handle equipment failures or other issues more easily. As for food, worms will eat almost anything that is biodegradable. You can use leaves, lawn clippings, manure, food scraps (vegetable or fruit), or even paper or cardboard.

Aquaponics Fish

Aquaponics it might be a new term many of us are hearing now a days, aquaponics is a combination of two words aquaculture (raising aquatic animals) and hydroponics (growing plants in water). Thus aquaponics is a method in which there is a close and long term biological relationship between the aquatic animals and plants were they both benefit from each other.

In aquaculture the fish wastes get accumulated in water and increase the toxicity of the water thus causing harm to the fishes. In hydroponics the plants require a continuous supply of nutrients which in much case are not organic. So to solve this, aquaponics came into play; the water from the aquaculture is fed into the hydroponic system. Here the plants feed on the fish waste as nutrients and thus purify the water and reduce the toxicity. This toxic free water is then pumped back into the aquaculture.

The fish plays an important role in the aquaponic system. The fish excretion is used as the nutrient for the plant growth. We need to consider certain factors while selecting fish for the aquaponic system not all fishes are suitable for aquaponics. The main rules or factors that need to be taken into consideration are:

Temperature: While selecting a fish first and foremost it should be able to survive in the temperature of your area. Fishes are mainly cold blooded animals, that means they don't regulate their body temperature instead they depend on the external environment they are in to regulate their body temperature. So climate temperature is essential while for fish selection.

Warm water fishes thrive in temperature >65 °F and < 85 °F. Common warm water fishes are Tilapia, catfish, Bass etc. Tilapia will die if the temperature falls below 50 °F. If your climate is more to the colder side then these fishes are not an ideal choice for aquaponics.

Cold water fishes thrive in temperature range of 50 °F to 60 °F. Common cold water fish is the various trout species. The best one to use in aquaponics is rainbow trout.

Easy availability: Next important matter to keep in mind while selecting fish is the ease of availability of the fish in your location. For example Barramundi is not easily available in US but it's readily available in Australia.

So the ease of acquiring fingerlings also matters while selecting fish for aquaponics. Now Tilapia is the most commonly used fish for aquaponics as they are readily available in most part of the world. China is the largest producers of Tilapia.

This is another important factor to consider while starting aquaponics, not all fishes are legal to be purchased and grown in certain locations. So we have to first check and make sure that the fish that you intend to use in aquaponics is legal in your location.

Even though Tilapia is considered the common fish for aquaponics they are considered as pest by many countries due to the high propagation rate and also this over growth can lead to cause of diseases which has resulted in ban of this species in many countries.

Here is a list of few fishes that are mainly used in aquaponic systems. A general comparison is given based on their temperature requirement, availability and legality.

Tilapia

Temperature: Warm water fish, tolerable temperature 65 °F to 85 °F. The aquaponic systems are usually kept at 73 degree Fahrenheit to accommodate the plants.

Availability: They are hardy fish, survival rate is high. They can withstand pH changes, ammonia level change and eat any feed given. The Nile and Mozambique Tilapia are the most favoured varieties, as their growth rate and breeding rate are high.

Legality: Legal in most part of the world. They are considered as invasive species in US, Australia and South Africa. It is considered illegal in some part of the world, so before buying check with your state fishery department.

Trout

Temperature: These are cold water fish; ideal temperature required for them is 58oF – 68oF. They survive in cold weather. The most desirable species of Trout for aquaponics is Rainbow Trout. They are also popular aquaponic species to be grown in garages and basements. Trout can't handle dirty water unlike Tilapia.

Availability: Due to their cold water habitat they are not readily available in the warmer climate regions. They are found in the natural habitats of North America, Northern Asia and Europe.

Legality: Even though they are not suitable to be raised in many parts of the world, there is a ban on trout farming in many countries. For example in New Zealand Trout farming is banned as they fear that this would put the recreational trout fishery at risk.

Perch

Temperature: Ideal temperature for perch is 70 to 82 degree Fahrenheit. They thrive in warm waters, but they can with stand water temperature up to 50 degree Fahrenheit.

Availability: they are edible, hardy and adaptable aquaponic fishes. Popular varieties of perch are silver, yellow and jade. They have a wide distribution throughout the world. Jade perch is readily available in Australia but difficult to find in U.S. Yellow perch is most commonly used in aquaponics as they require only moderate temperature and can withstand wide pH range.

Legality: Perch farming is legal in most part of the World and Perch fishing is a recreational sport for many countries.

Catfish

Temperature: Ideal temperature for catfish is 78 to 86 degree Fahrenheit. They thrive in warm waters and they are sensitive to sudden water temperature changes.

Availability: They are bottom-feeding aquaponic fishes. They require pristine water and pH also should be ideal. They are fresh water fishes and live inland or in coastal waters of all continents except Antarctica.

Legality: Catfish farming is legal in most part of the World.

Barramundi

Temperature: Ideal temperature is 78 to 83 degree Fahrenheit. They thrive in warm waters and are fast growers.

Availability: They are edible aquaponic fishes. Popular in Australia, but is raised in Singapore, Saudi Arabia, India, Indonesia etc. They can be kept in both freshwater and salt water. The large fishes will eat the small fishes so breeding of fingerlings are difficult. They need good quality water and lots of dissolved oxygen.

Legality: Barramundi farming is legal in most part of the World. Their meat is highly priced and rich in omega 3 fatty acid.

Bass

Temperature: Ideal temperature for Bass is 75 to 85 degree Fahrenheit. They thrive in warm waters, but they are not hardy and adaptable as the other species. They can tolerate low water temperatures.

Availability: Top aquaponic bass varieties are largemouth, smallmouth and striped bass. They are readily available as they are considered as sport fish in many places. Hybrid striped bass are well suited for aquaponics as they are resilient to temperature changes.

Legality: They are legal in many part of the world, but are being considered as invasive species in many places like Canada, Northern America.

Crustaceans: Prawn, Lobster, Oyster and Mussels

Temperature: deal temperature for prawns are 82 to 88 °F, lobster 71- 76 °F, oyster 75 – 79 °F. They can be grown alongside other fisher or in separately. For temperate climate Red claw is a fast growing variety of crayfish and for cooler climate Yabbies and Marron are best suited.

Availability: Crustaceans are readily available in all part of the world. They can be grown hassle free as they are mostly bottom feeders or filter feeders. They help to keep the aquaponic system clean.

Legality: Crustacean farming is legal in most part of the World and they are a delicacy in many countries.

Koi

Availability: They have gained popularity as ornamental fish in many part of the world. They are readily available but are a bit on the expensive side. Thy have high resilience to parasites and diseases.

Legality: Koi farming is legal in most part of the World and mainly they are farmed for ornamental purpose. They have long lifespan of about 60 years.

Temperature: Ideal temperature for Koi is 65 to 75 degree Fahrenheit. These are the most popular ornamental fish.

Goldfish

Temperature: Ideal temperature for goldfish is 68 to 75 degree Fahrenheit. They are hardy and adaptable to various water conditions. Rapid temperature changes can be fatal for them.

Legality: Goldfish farming is legal in all part of the World and mainly they are farmed for ornamental purpose. They have long lifespan of about 15 years. They produce lot of beneficial nutrient for aquaponic system.

Availability: They have gained much popularity as ornamental fish in all part of the world. They are small in size and provide lot of nutrients. Due to small size they are susceptible to parasites.

Carp

Temperature: Ideal temperature for carp is 80 to 82 degree Fahrenheit. They thrive in warm waters. They are oily freshwater fish. They are popular in aquaculture.

Availability: They are largely available in all parts of the world. However their demand have reduced in Western Europe due to the increased farming of species like trout, salmon etc.

Legality: They are also legal in many part of the world. They have good reproductive capabilities and can survive in most environments.

Aquaponic Vegetables

An aquaponics system can be used to grow nearly any type of plants, vegetables, fruits, herbs, and even flowering plants. However, the question that might come to mind is: What are the best plants to grow in aquaponics?

Aquaponics Gardening

Almost everyone has at least some understanding of how vegetable gardening works. In a traditional method of gardening, you plant your seeds, water and add fertilizer occasionally, remove weeds or insects, and wait to see whether your plant will flourish or not.

Aquaponics works differently from this method because an aquaponics system works by creating a nitrogen cycle. In this cycle, the three main elements, the fish, plants, and bacteria, shared the water. In the fish tank, fish produced high ammonia wastes. The pump carries this wastes into the grow bed, where the bacteria convert these wastes into nitrites and then into nitrates.

These nitrates become food for the plants to grow, and in return, the plant roots clean the water in the grow beds before it returns to the fish tank for the fish to live. These are a continuous cycle where the bacteria convert fish wastes into plant's food, and plants clean the water for the fish.

Things to Consider in Choosing the Best Plants for Aquaponics System

Almost all plants can be grown in an aquaponic system. However, different plants have different needs and thrive under different conditions. Before you start choosing which plants to grow in your aquaponics system, it is important to consider these things.

Type of System to use

This could be a media-based, NFT, raft aquaponics system. These will determine the type of root structure that plants tend to have. Plants with no root structure will grow well in floating rafts; while root vegetables grow well in grow beds.

Consider the Need of your Plants

As a rule of thumb, the more similar your plant's needs to the needs of the fish in your system, the better they will grow. Select plants and fish that have similar needs as far as temperature and pH level are concerned because the closer they match the more success you will have.

Environment

The amount of sunlight, temperature, and rain are all important in growing a healthy plant. If you decide to grow outdoors, choose a variety of vegetables that will grow best in your climate.

You could always use a greenhouse or grow indoors, but you need to find plants that do well inside with limited or artificial sunlight. It is best to choose a plant that thrives in your climate, as this will also lower the electricity cost in maintaining your system.

Here are some crops that grow in a warm or colder climate. You can either follow the climate requirements for these plants or use a greenhouse to regulate the temperature and stay with the few aquaponics plants you prefer.

Cold crops	Warm crops
Beets	Artichoke
Brussels sprouts	Cardon
Cabbage	Chives
Carrots	Garlic
Cauliflower	Parsley
Herbs	Parsnip
Horseradish	Shallot
Onions	Sludge
Potatoes	
Radicchio	
Sweet potatoes	
Winter squash	

Space

This refers to the amount of space available in your system. Some plants require more space than others. So carefully consider the height and width needed for the plants to grow, and how many plants could fit in your system.

The Number of Fish

It is important to consider the fish to plant ratio in your system. If you plan on growing much fish, you will need more plants to absorb all the nutrients from the fish waste.

What you want to Grow

Aquaponics gardening tends to go better when you are growing plants that you want to eat or use. Thinking about what you want to achieve in your system will help you make a better decision on what plant to grow in your aquaponics system.

Best Plants to Grow in a New Aquaponics System

Some plants have low nutrient needs and grow well in a newly built aquaponic system, making them the best "starter" plants for newly built aquaponic systems. It is best to start a new aquaponic system with fast-growing robust plants with low nutrient demand. Some examples are leafy green vegetables; these will usually grow well in a newly built aquaponics system.

Lettuce

The easiest and one of the most productive leafy greens in an aquaponics system is lettuce. This is because leafy lettuce thrives in water, which has a temperature of between 70 and 74-76 °F making it great to pair with tilapia. Lettuce has minimal nutritional requirements and can thrive in many environments making it a great confidence building plant to grow in a new aquaponics system.

Watercress

Watercress is one of the best plants for aquaponics because it is a hardy vegetable and grows quickly. Watercress is easier to maintain than other plants due to its low nutritional demand and can exist alongside many different types of fish.

Cucumber

The cucumber is a water-based plant, and an aquaponics system gives them everything they need to flourish. Cucumbers grow best in water with a pH between 5.5 and 7 and grow best in media-based aquaponics systems.

Kale

Kale is not a very nutritious demanding plant to grow, but it is a very nutritious plant. Kale can match and exist alongside many types of fish that you may have in your system. Kale is not attractive to pests making it one of the best plants to grow in aquaponics.

Swiss Chard

Swiss chard is a great plant to grow in your aquaponics system. Swiss chard is frost resistant, making it an ideal plant during the winter season. The nutrients required to grow Swiss chard are moderate, using mostly nitrates for leaf growth. Swiss chard grows best in waters with a pH level between 6.6 to 7.0, which is ideal for the plant and many fish.

Best Plants to Grow in an Established Aquaponics System

Different plants have different nutritional needs, and your aquaponic system needs to be established to support plants with higher nutrient demands. A well-established system contains all the beneficial bacteria necessary. A well-established aquaponics system is a system that has been running for at least 6 months and up to 3 years.

These are the plants that grow well in an aquaponics system but require higher nutrient needs. These are also the best plants to grow in established aquaponic systems.

Tomatoes

Tomato plants do exceptionally well in aquaponics systems. It like warm weather but also need plenty of nutrient-rich water. They are an excellent option for an aquaponics system where the environment can be controlled.

Cauliflower

Cauliflower plants need very little maintenance. They are also resistant to most bugs and diseases. It is a water-based plant that thrives in aquaponics systems.

Cabbage

Cabbage needs very little maintenance, which makes it one of the best plants in an aquaponics system. They thrive in water with a pH level between 6.2 and 6.6.

Peppers

Peppers can be challenging to grow in traditional gardening, but it is one of the best plants to grow in your aquaponics system. Peppers are very particular about the water that they consume and they require plenty of sunshine. Growing peppers using aquaponics allows you to monitor, and control your nutrients in the water and lets you create the perfect environment for your pepper plants.

Strawberries

Strawberries need very little space to grow and are easy to cultivate in an aquaponics system. They are one of the best plants to grow in aquaponics because of the constant supply of nutrient-rich water and the ability to monitor and control the system environment. Another benefit is that you can grow them all year round in an indoor system.

Crops or vegetables are not the only plants that you can grow in aquaponics; in fact, flowers are a good choice as they are safe from weeds and most pests in an aquaponics system. Here are some great flowers to grow using aquaponics.

Marigolds

Aside from giving a wonderful burst of yellow color to your aquaponics garden, they have strong odor that confuses and drive pests away from your plants.

Water Hyacinth

Many people consider water hyacinth to be invasive. However, the water hyacinth is excellent in cleaning and purifying the water for your fish, which makes it perfect for fish that are sensitive to water changes.

Aquaponics Systems

Since aquaponics uses basically the same systems as hydroponics, there aren't many differences in how the system works, except for the added fish in the water tank(s). Drip irrigation, flood and drain, deep culture or water submerged roots, and nutrient film technique are highly compatible and customizable to merge with growing fish. Primary methods of aquaponic growing widely in use today:

- Deep Water Culture: (DWC) or raft based growing uses a foam raft that is floating in a channel filled with fish effluent water that has been filtered to remove solid wastes. Plants are placed in holes in the raft and the roots dangle freely in the water. This method is most appropriate for growing salad greens and other fast-growing, relatively low-nutrient plants. It is also most commonly used in larger commercial-scale systems.
- Media-based Aquaponics: Media growing involves growing plants in inert planting media such as expanded clay pellets or shale. The media provides both the biological filtration (conversion of ammonia to nitrates) and mechanical filtration (removal of solid wastes) in the same system. Media-based systems are great for home and hobby scale systems so you can grow a wide variety of crops. In particular, large fruiting plants do really well in addition to leafy greens, herbs and other varieties. Examples of media-based systems are the Harmony (pictured), the Aqua Urban 60 gallon system, and the AquaBundance system.
- Nutrient Film Technique (NFT): NFT systems work by flowing nutrient-rich water through a narrow trough, such as a PVC pipe. Plants are placed in holes drilled in this pipe, and the roots dangle freely in this stream of water. This method of growing works very well for plants that needs little support, such as strawberries (pictured) and other herbs. NFT is also a great way to utilize unused space because they can be hung from ceilings above other growing areas.
- Vertical Aquaponics: One of the greatest aspects of aquaponics is its ability to grow an incredible amount of food in a very small area. No method does this well than vertical aquaponics. Plants are stacked on top of each other in tower systems such as the AquaVertica. Water flows in through the top of the tower, and flows through a wicking material that the plants roots absorb water and nutrients from. The water then falls into a trough or directly into the fish tank. This form of agriculture makes the most of each square foot of space, and works very well with leafy greens, strawberries, and other crops that do not require support to grow.

Nutrient Cycling in Aquaponics Systems

Aquaponic systems offer various advantages when it comes to producing food in an innovative and sustainable way. Besides the synergistic effects of increased aerial CO_2 concentration for greenhouse crops and decreased total heat energy consumption when cultivating fish and crops in the same space, aquaponics has two main advantages for nutrient cycling. First, the combination of a recirculating aquaculture system with hydroponic production avoids the discharge of aquaculture effluents enriched in dissolved nitrogen and phosphorus into already polluted groundwater, and second, it allows for the fertilisation of the soilless crops with what can be considered an organic solution instead of using fertilizers of mineral origin made from depleting natural resources. Furthermore, aquaponics yields comparable plant growth as compared with conventional hydroponics despite the lower concentrations of most nutrients in the aquaculture water, and production can be even better than in soil. Increased CO_2 concentrations in the aerial environment and changes in the biomes of the root zone are thought to be main reasons for this. In addition, the mineral content and the nutritional quality of tomatoes grown aquaponically have been reported to be equivalent or superior to the mineral content of conventionally grown ones.

Despite having two attractive assets (i.e. the recycling of aquaculture effluents and the use of organic fertilizers), the use of aquaculture effluents increases the challenge of monitoring the nutrients within the solution. Indeed, it is harder to control the composition of a solution where the nutrients originate from a biological degradation of organic matter than to follow the evolution of the nutrients' concentration in a precisely dosed hydroponic solution based on mineral compounds. Moreover, a plant's nutritional needs vary during the growth period in accordance with physiological stages, and it is necessary to meet these needs to maximize yields. In order to recycle aquaculture effluents to produce plant biomass, it is necessary to optimize the recycling rates of phosphorus and nitrogen. Several factors can influence this, such as the fish species, fish density, water temperature, the type of plants and the microbial community (ibid.). Therefore, it is of prime importance to understand the functioning of the nutrient cycles in aquaponics.

Origin of Nutrients

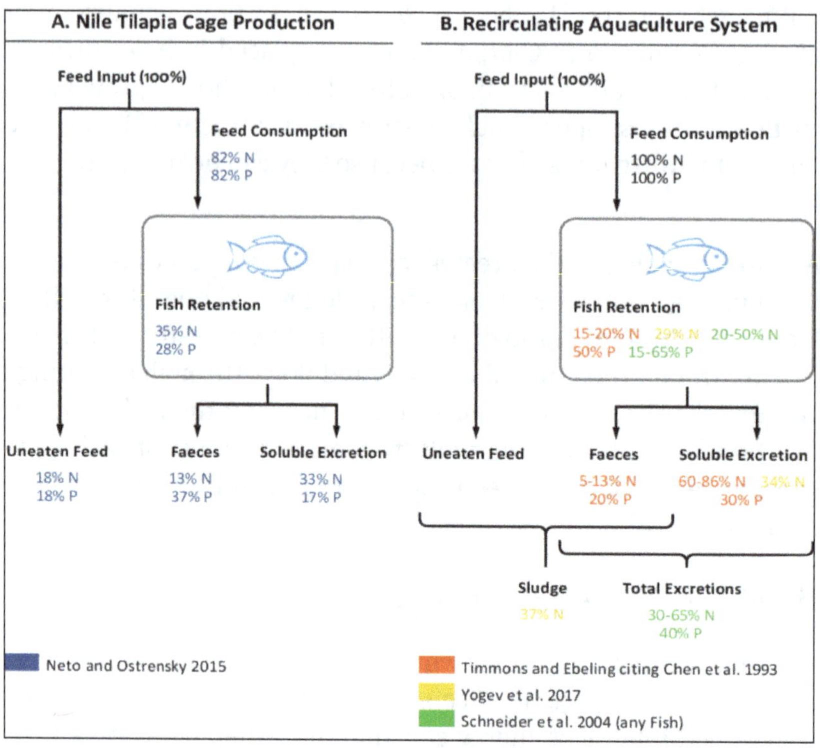

The major sources of nutrients in an aquaponic system are the fish feed and the water added (containing Mg, Ca, S) into the system. With respect to fish feed, there are two main types: fishmeal-based and plant-based feed. Fishmeal is the classic type of feed used in aquaculture where lipids and proteins rely on fish meal and fish oil. However, for some time now, concerns regarding the sustainability of such feed have been raised and attention drawn towards plant-based diets. A meta-analysis conducted by Hua and Bureau revealed that the use of plant proteins in fish feed can influence the growth of fish if incorporated in high proportions. Indeed, plant proteins can have an impact on the digestibility and levels of anti-nutritional factors of the feed. In particular, phosphorus originating from plants and thus, in the form of phytates do not benefit, for example, salmon, trout and several others fish species. It is not surprising that this observation is highly dependent on the fish species and on the quality of the ingredients. However, little is known of the impact of varying fish feed composition on crop yields? Classical fish feed is composed of 6–8 macro ingredients and contains 6–8% organic nitrogen, 1.2% organic phosphorus and 40–45% organic carbon with around 25% protein for herbivorous or omnivorous fish and around 55% protein for carnivorous fish. Lipids can be fish or plant based as well.

Once fish feed is added into the system, a substantial part of it is eaten by the fish and either used for growth and metabolism or excreted as soluble and solid faeces, while the rest of the given feed decays in the tanks. In this case, the feed leftovers and metabolic products are partly dissolved in the aquaponic water, thus enabling the plants to uptake nutrients directly from the aquaponic solution. In most cultivation systems, nutrients can be added to complement the aquaponic solution and ensure a better matching with the plants' needs. Indeed, even when the system is coupled, it is possible to add iron or potassium (which are often lacking) without harming the fish.

Fish Feed Leftovers and Fish Faeces

Ideally, all the given feed should be consumed by the fish. However, a small part (less than 5% is often left to decompose in the system and contributes to the nutrient load of the water, thus consuming dissolved oxygen and releasing carbon dioxide and ammonia, amongst other things. The composition of fish feed leftovers depends on the composition of the feed. Logically enough, the composition of fish faeces depends on the fish's diet which also has an impact on the water quality. However, the nutrient retention in fish biomass is highly dependent on fish species, feeding levels; feed composition, fish size and system temperature. At higher temperatures, for example, fish metabolism is accelerated and thus results in more nutrients contained in the solid fraction of the faeces. The proportion of excreted nutrients also depends on the quality and digestibility of the diet. The digestibility of the fish feed, the size of the faeces and the settling ratio should be carefully considered to ensure a good balance in the system and to maximize crop yields. Indeed, while it is a priority that fish feed should carefully be chosen to suit fish needs, the feed components could also be selected to suit plant's requirements when it makes no difference to fish.

Microbiological Processes

Solubilization

Solubilization consists of the breaking down of the complex organic molecules composing fish waste and feed leftovers into nutrients in the form of ionic minerals which plants can absorb. In both aquaculture and aquaponics, solubilization is conducted mainly by heterotrophic bacteria which have not yet been fully identified. Some studies have started deciphering the complexity of these bacteria communities. In current aquaculture, the most commonly observed bacteria are Rhizobium sp., Flavobacterium sp., Sphingobacterium sp., Comamonas sp., Acinetobacter sp.,

Aeromonas sp. and Pseudomonas sp. An example of the major role of bacteria in aquaponics could be the transformation of insoluble phytates into phosphorus (P) made available for plant uptake through the production of phytases which are particularly present in γ-proteobacteria. (More research needs to be done in this area). Other nutrients than P can also be trapped as solids and evacuated from the system with the sludge. Efforts are thus being made to remineralise this sludge with UASB-EGSB reactors in order to reinject nutrients into the aquaponic system. Furthermore, different minerals are not released at the same rate, depending on the composition of the feed, thus leading to more complicated monitoring of their concentration in the aquaponic solution.

Nitrification

The main nitrogen source in an aquaponic system is the fish feed and the proteins it contains. Ideally, 100% of this feed should be eaten by the fish. However, is has been observed that fish only use about 30% of the nitrogen contained in the given feed. The ingested feed is partly used for assimilation and metabolism, while the rest is excreted either through the gills or as urine and faeces. The nitrogen which is excreted through the gills is mainly in the form of ammonia, NH3, while urine and faeces are composed of organic nitrogen which is transformed into ammonia by proteases and deaminase. In general, the fish excrete nitrogen under the form of TAN, i.e. NH_3 and NH_4^+. The balance between NH_3 and NH_4^+ depends mostly on the pH and temperature. Ammonia is the major waste produced by fish catabolism of the feed proteins.

Nitrification is a two-step process during which the ammonia NH_3 or ammonium NH_4^+ excreted by the fish is transformed first into nitrite NO_2 and then into nitrate NO_3 by specific aerobic chemosynthetic autotrophic bacteria. A high availability of dissolved oxygen is required as nitrification consumes oxygen. The first step of this transformation is carried out by ammonia-oxidizing bacteria (AOB) such as Nitrosomonas, Nitrosococcus, Nitrospira, Nitrosolobus and Nitrosovibrio. The second step is conducted by nitrite-oxidizing bacteria (NOB) such as Nitrobacter, Nitrococcus, Nitrospira and Nitrospina. Nitrospira is currently deduced to be a complete nitrifier, i.e. to be involved in the production of both nitrite and nitrate. The same bacteria can be found both in aquaculture and aquaponic systems. These bacteria are mainly found in biofilms fixed to the media composing the biofilter but can also be observed in the other compartments of the system.

Nitrification is of prime importance in aquaponics as ammonia and nitrite are quite toxic for fish: 0.02–0.07 mg/L of ammonia–nitrogen are sufficient to observe damage in warm water fish, and nitrite–nitrogen should be kept under 1 mg/L. Ammonia affects the central nervous system of the fish, while nitrite induces problems with oxygen fixation. Nitrate– nitrogen is, however, tolerated by the fish up to 150–300 mg/L.

Nitrification mostly takes place in biofilters. Therefore, when starting a system, it is recommended to run the system without fish in order to allow the slowly growing population of nitrifying bacteria to establish. It is also necessary to avoid, as far as possible, the presence of organic matter in the biofilters in order to prevent the growth of highly competitive heterotrophic bacteria. Alternatively, commercial mixes of nitrifying bacteria can be added to the system, prior to stocking, to hasten the colonization process. Nevertheless, small aquaponic systems without biofilter also exist. In these systems, nitrifying bacteria form biofilms of the available surfaces (e.g. hydroponic compartment walls, inert media when using the media bed technique).

Mass Balance

The functioning of aquaponic systems is based on a dynamic equilibrium of the nutrient cycles. It is therefore necessary to understand these cycles in order to optimize the management of the systems. Plants growing hydroponically have specific requirements, which should be met during their various growing stages. Therefore, nutrient concentrations in the different compartments of the system must be closely monitored, and nutrients should be supplemented to prevent deficiencies either in the system water or via foliar application. In some cases, supplementing an aquaponic solution with mineral nutrients in order to reach the same nutrient concentrations as in hydroponics could lead to higher yields than those achieved in hydroponics. The first step to take towards a balanced system is the correct design and relative sizing of the compartments. If the hydroponic compartment is too small compared to the fish tanks, then the nutrients will accumulate in the water and could reach toxic levels. The feed rate ratio (i.e. the amount of fish feed in the system based on the plant-growing surface and the plant type) is often used for the first sizing of the system. However, it is not possible to reach a plant/ fish ratio which will enable an optimal match of plants' needs if only fish feed is used as an input. To make sure that the system is well balanced and functions properly, monitoring methods are usually based on the nitrogen cycle (Cerozi and Fitzsimmons 2017; Somerville et al. 2014), but to ensure optimal functioning of the system, it is necessary to monitor more closely the balance of the other macronutrients (N, P, K, Ca, Mg, S) and micronutrients (Fe, Zn, B, Mn, Mo, Cu) as well. Recent studies have started tackling this topic. Schmautz et al. compared the impact of three different hydroponic layouts (i.e. nutrient film technique (NFT), floating raft and drip irrigation) on the nutrient uptake of aquaponic tomatoes. Drip irrigation was the system which produced slightly better yields with tomatoes. The mineral content of the fruits (P, K, Ca, and Mg) was equivalent to the conventional values even though the iron and zinc contents were higher. The leaves however had lower levels of P, K, S, Ca, Mg, Fe, Cu and Zn than in conventional agriculture. Delaide et al. followed the cycles of macro- and micronutrients in a coupled aquaponic system. They observed that K, P, Fe, Cu, Zn, Mn and Mo were lacking in their aquaponic solution, while N, Ca, B and Na were quickly accumulated. Graber and Junge noted that their aquaponic solution contained three times less nitrogen and ten times less phosphorus than a hydroponic solution. As for potassium (K), it was 45 times lower compared to hydroponics. Nevertheless, they obtained yields as similar yields even though the quality was poorer due to a lack of potassium (K).

Factors Influencing the Nutrient Cycles

Light intensity, root zone temperature, air temperature, nutrient availability, growth stage and growth rate all influence a plant's nutrient uptake. Experiments conducted by Schmautz et al. and Lennard and Leonard showed that the hydroponic method could also play a role in a plant's nutrient uptake capacity, and it is therefore necessary to match the growing system to the type of vegetables being grown. NFT and DWC (deep water culture – raft) are thus suitable for leafy greens, whereas drip irrigation on rock wool slabs is more suitable for fruity vegetables.

Macronutrient Cycles

Carbon (C)

Carbon is provided to the fish via the feed and to the plants via CO_2 fixation. Fish can use 22%

of the carbon contained in the fish feed for biomass increase and metabolism. The rest of the ingested carbon is either expirated under the form of CO_2 (52%) or excreted in a dissolved (0.7–3%) and solid (25%) form. The expirated CO_2 can be used by plants for their own carbon source as well. The uneaten part of the feed carbon is left to decompose in the system. The type of carbohydrates found in fish feed (e.g. starch or non-starch polysaccharides) can also influence the digestibility of the feed and the biodegradability of the waste in an aquaculture or aquaponic system.

Nitrogen (N)

Nitrogen is absorbed by the plants either in the nitrate or ammonium form depending on the concentration and plant's physiology. Associations between plants and microorganisms should not be overlooked as plants affect the presence of the microorganisms in aquaponics, and microorganisms can play a significant role in the nitrogen uptake capacity of plants. The uptake of nitrogen by plants is affected by the ambient carbon dioxide concentration.

Phosphorus (P)

Phosphorus is one of the essential elements for plant growth and can be absorbed under its ionic orthophosphate form (H_2PO_4, HPO_4^{2-}, PO_4^{3-}). Little is known about the dynamics of phosphorus in aquaponics. The main input of phosphorus in the system is the fish feed, and in un-supplemented systems, phosphorus tends to be limiting and thus can impede plant growth, fish can use up to 15% of the phosphorus contained in the feed. In a system growing lettuce, Cerozi and Fitzsimmons noticed that the amount of phosphorus provided by the fish feed can be sufficient or insufficient depending on the growth stage. Up to 100% of phosphorus present in the fish water can be recycled in the plant biomass, depending on the design of the system. Graber and Junge observed a 50% recycling, while Schmautz et al. reported that 32% of the phosphorus could be found in the fruit and 28% in the leaves. The solubility of phosphorus depends on the pH, and a higher pH will foster the precipitation of phosphorus, thus rendering it unavailable for the plants. Phosphorus can precipitate as struvite (magnesium ammonium phosphate) and/or hydroxyapatite. These insoluble complexes are removed via solid fish sludge from the system. Schneider et al. (2004) reported that 30–65% of the phosphorus contained in the fish feed remains unavailable to plants as it is fixed in the solid excretions which are then removed through mechanical filtration. Yogev et al. estimated that this loss can be up to 85%. One option to prevent this massive loss of P via solid sludge is to add a digestion compartment to the aquaponic system. During aerobic or anaerobic digestion, the P is released into the digestate and could be re-introduced into the circulating water.

Potassium (K)

Delaide et al. found that the major source of K in their system was the fish feed. Fish can use up to 7% of the K contained in the fish feed. However, potassium is not necessary for fish which leads to a low potassium composition of the fish feed and to even lower levels of potassium available for the plants. To supply potassium, a KOH pH buffer is often used as the pH often decreases in aquaponics due to nitrification. In an aquaponic system planted with tomatoes, potassium accumulated mainly in the fruits.

Magnesium (Mg), Calcium (Ca) and Sulphur (S)

The main source for Mg, Ca and S is tap water which facilitates the absorption by the plants as the nutrients are already available. Calcium is however present in insufficient levels in aquaponics and is added under the form of calcium hydroxide $Ca(OH)_2$. According to Rafiee and Saad, fish can use on average 26.8% of the calcium and 20.3% of the magnesium present in the feed. Sulphur is often at low levels in aquaponic systems.

Micronutrient Cycles

Iron (Fe), manganese (Mn) and zinc (Zn) derive mainly from the fish feed, while boron (B) and copper (Cu) derive from the tap water. In aquaponics, key micronutrients are often present but at too low levels and supplementation from external sources of nutrients is then necessary. Iron deficiencies occur very often in aquaponics, mostly because of the non-availability of the ferric ion form. This deficiency can be solved by the use of bacterial siderophore (i.e. organic iron-chelating compounds) produced by genera such as Bacillus or Pseudomonas or by iron supplementation with chemical chelated iron to avoid iron precipitation.

Nutrient Losses

Reducing nutrient loss is a constant challenge facing aquaponics practitioners. Nutrient loss occurs in several ways, e.g. the settlement of the sludge (37% of faeces and 18% of uneaten feed), water losses, denitrification, ammonia volatilization, etc. As an example, Rafiee and Saad note that 24% of the iron, 86% of the manganese, 47% of the zinc, 22% of the copper, 16% of the calcium, 89% of the magnesium, 6% of the nitrogen, 6% of the potassium and 18% of the phosphorus contained in the fish feed were contained in the sludge. The sludge can hold up to 40% of the nutrients present in the feed input.

Denitrification can lead to a loss of 25–60% of the nitrogen. Denitrification is also linked to anoxic conditions and low carbon levels and is responsible for the transformation of nitrate into nitrite, nitric oxide (NO), nitrous oxide (N_2O) and eventually nitrogen gas (N_2) with flows into the atmosphere (ibid.). Denitrification is conducted by several facultative heterotrophic bacteria such as Achromobacter, Aerobacter, Acinetobacter, Bacillus, Brevibacterium, Flavobacterium, Pseudomonas, Proteus and Micrococcus sp. Some bacteria can perform both nitrification and denitrification if dissolved oxygen levels are below 0.3 mg/L. The loss of nitrogen can also occur via anaerobic ammonium oxidation (ANAMMOX), i.e. the oxidation of ammonium into dinitrogen gas in the presence of nitrite.

Another important loss of nitrogen which should be available for the plants is the consumption of the nitrogen by the heterotrophic aerobic bacteria present in the aquaponic systems. Indeed, the nitrogen used by these bacteria is lost to nitrifying bacteria, and nitrification is thus impeded. These bacteria are particularly present when the C/N ratio increases as they are more competitive and more able to colonize the media than the autotrophic nitrifying bacteria.

Nutrient Balance Systems Dynamics

The nutrient concentration of the two major subsystems in an aquaponic system, i.e. the fish tanks (aquaculture) and the hydroponic solution, needs to be balanced for each of their needs. In closed aquaponic systems, nutrients are transported from fish to plants more or less directly through filters (usually a drum filter or a settler and then a biofilter) for nitrification. However, the nutrient needs of crops and the supplied nutrients from the aquaponic subsystem are not in balance. In multi-loop and decoupled systems, it is easier to provide optimal conditions for both, fish and plant sections. Through modeling of the systems, the optimal size of the hydroponic area to the fish tanks, biofilters and other equipment can be calculated. This is particularly important with decoupled multi-loop systems which comprise various types of equipment. For example, UASB (up flow anaerobic sludge blanket) or desalination units need to be sized carefully. The basic mismatch of nutrients supplied by the fish environment and the needs of the crops needs to be rectified and balanced. For the purpose of up-concentrating the nutrients, this approach, however, only solves part of the problem, as the perfectly balanced system is driven by a non-dynamic evaporation rate achievable only in closed chambers and perfectly working plants. The reality, however, is that the evapotranspiration of the crop (ET_c) in greenhouse based aquaponics systems is highly dependent on multiple factors such as physical climate and biological variables. ET_c is calculated per area of the ground surface covered by the crop and is calculated for different levels in the canopy (z) by integrating irradiative net fluxes, boundary layer resistance, stomata resistance and the vapour pressure deficit, in the canopy using the Penman– Monteith equation. This equation, nevertheless, only calculates the water flux through the crop. Nutrient uptake can either be calculated simply by assuming that all diluted nutrients in the water are taken up by the crop. In reality though, uptake of nutrients is a highly complicated matter. Different nutrients have different states, changing with parameters such as pH. Meanwhile nutrient availability to the plants strongly depends on pH and the relationships of nutrients to each other (e.g. K/Ca availability). In addition, the micro biome in the root zone plays an important role which is not yet implemented in models although some models differentiate between phloem and xylem pathways. Thus, the vast amount of nutrients is not modeled in detail for aquaponic nutrient balancing and sizing of systems. The easiest way to estimate nutrient uptake is the assumption that nutrients are taken up/absorbed as dissolved in irrigation water and apply to the above explained Etc. calculation approach and assuming that no element-specific chemical, biological or physical resistances exist. Consequentially, to maintain equilibrium, all nutrients taken up by the crop as contained in the nutrient solution need to be added back to the hydroponic system.

References

- Aquaponics-information, aquaponics-in-schools: aquaponics.com, Retrieved 10, April 2020
- What-is-aquaponics-biofilter: gogreenaquaponics.com, Retrieved 21, April 2020
- Hydroponics-advantages-and-disadvantages: dyna-gro.com, Retrieved 08, May 2020
- Aeroponic-technology-blessing-or-curse: ijert.org, Retrieved 27, May 2020
- Vermiponics-an-easy-hydroponics-alternative-system: thetyedyediguana.com, Retrieved 16, June 2020
- Best-fish-aquaponics: gardeningheavn.com, Retrieved 25, June 2020
- What-are-the-best-plants-for-aquaponics: gogreenaquaponics.com, Retrieved 09, June 2020
- What-is-aquaponics: theaquaponicsource.com, Retrieved 18, July 2020

Permissions

All chapters in this book are published with permission under the Creative Commons Attribution Share Alike License or equivalent. Every chapter published in this book has been scrutinized by our experts. Their significance has been extensively debated. The topics covered herein carry significant information for a comprehensive understanding. They may even be implemented as practical applications or may be referred to as a beginning point for further studies.

We would like to thank the editorial team for lending their expertise to make the book truly unique. They have played a crucial role in the development of this book. Without their invaluable contributions this book wouldn't have been possible. They have made vital efforts to compile up to date information on the varied aspects of this subject to make this book a valuable addition to the collection of many professionals and students.

This book was conceptualized with the vision of imparting up-to-date and integrated information in this field. To ensure the same, a matchless editorial board was set up. Every individual on the board went through rigorous rounds of assessment to prove their worth. After which they invested a large part of their time researching and compiling the most relevant data for our readers.

The editorial board has been involved in producing this book since its inception. They have spent rigorous hours researching and exploring the diverse topics which have resulted in the successful publishing of this book. They have passed on their knowledge of decades through this book. To expedite this challenging task, the publisher supported the team at every step. A small team of assistant editors was also appointed to further simplify the editing procedure and attain best results for the readers.

Apart from the editorial board, the designing team has also invested a significant amount of their time in understanding the subject and creating the most relevant covers. They scrutinized every image to scout for the most suitable representation of the subject and create an appropriate cover for the book.

The publishing team has been an ardent support to the editorial, designing and production team. Their endless efforts to recruit the best for this project, has resulted in the accomplishment of this book. They are a veteran in the field of academics and their pool of knowledge is as vast as their experience in printing. Their expertise and guidance has proved useful at every step. Their uncompromising quality standards have made this book an exceptional effort. Their encouragement from time to time has been an inspiration for everyone.

The publisher and the editorial board hope that this book will prove to be a valuable piece of knowledge for students, practitioners and scholars across the globe.

Index

www.ingramcontent.com/pod-product-compliance
Lightning Source LLC
LaVergne TN
LVHW080736210926
848485LV00003B/568